T0206230

# Systems Engineering Models

# Systems Innovation Series

*Series Editor:*
Adedeji B. Badiru
*Air Force Institute of Technology (AFIT) – Dayton, Ohio*

Systems Innovation refers to all aspects of developing and deploying new technology, methodology, techniques, and best practices in advancing industrial production and economic development. This entails such topics as product design and development, entrepreneurship, global trade, environmental consciousness, operations and logistics, introduction and management of technology, collaborative system design, and product commercialization. Industrial innovation suggests breaking away from the traditional approaches to industrial production. It encourages the marriage of systems science, management principles, and technology implementation. Particular focus will be the impact of modern technology on industrial development and industrialization approaches, particularly for developing economics. The series will also cover how emerging technologies and entrepreneurship are essential for economic development and society advancement.

Defense Innovation Handbook:
Guidelines, Strategies, and Techniques
*Edited by Adedeji B. Badiru and Cassie Barlow*

Productivity Theory for Industrial Engineering
*Ryspek Usubamatov*

Quality Management in Construction Projects
*Abdul Razzak Rumane*

Company Success in Manufacturing Organizations:
A Holistic Systems Approach
*Ana M. Ferreras and Lesia L. Crumpton-Young*

Introduction to Industrial Engineering
*Avraham Shtub and Yuval Cohen*

Design for Profitability: Guidelines to Cost Effectively Manage the Development Process of Complex Products
*Salah Ahmed Mohamed Elmoselhy*

For more information about this series, please visit: https://www.crcpress.com/Systems-Innovation-Book-Series/book-series/CRCSYSINNOV

# Systems Engineering Models

## Theory, Methods, and Applications

Adedeji B. Badiru

CRC Press is an imprint of the
Taylor & Francis Group, an **informa** business

CRC Press
Taylor & Francis Group
6000 Broken Sound Parkway NW, Suite 300
Boca Raton, FL 33487-2742

First issued in paperback 2020



ISBN-13: 978-1-138-57761-9 (hbk)
ISBN-13: 978-0-367-78013-5 (pbk)



**Library of Congress Cataloging-in-Publication Data**

Names: Badiru, Adedeji Bodunde, 1952- author.
Title: Systems engineering models : theory, methods, and applications / authored by Adedeji B. Badiru.
Description: Boca Raton : Taylor & Francis, a CRC title, part of the Taylor & Francis imprint, a member of the Taylor & Francis Group, the academic division of T&F Informa, plc, 2019. | Series: Systems innovation series | Includes bibliographical references.
Identifiers: LCCN 2018053760 | ISBN 9781138577619 (hardback : acid-free paper) | ISBN 9781351266529 (e-book)
Subjects: LCSH: Simulation methods. | Systems engineering.
Classification: LCC T57.62 .B33 2019 | DDC 620/.0042—dc23
LC record available at https://lccn.loc.gov/2018053760

**Visit the Taylor & Francis Web site at**
**http://www.taylorandfrancis.com**

**and the CRC Press Web site at**
**http://www.crcpress.com**

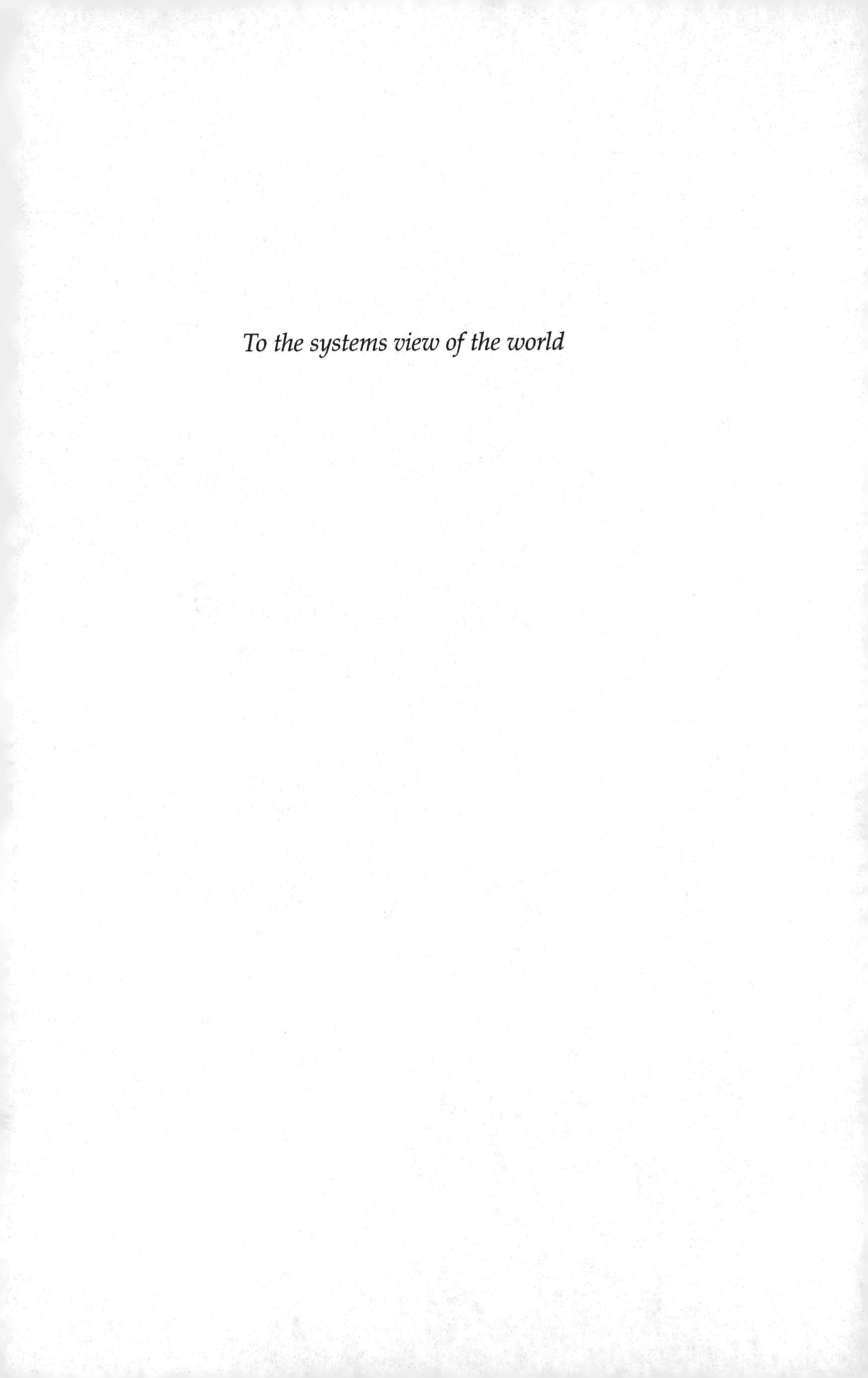

*To the systems view of the world*

# Contents

# Preface

There is a need for all engineers and nonengineers to understand the interdisciplinary and cross-functional nature of systems engineering, and the benefits of following rigorous systems engineering processes in organizational pursuits. This book provides the knowledge and understanding necessary to achieve a basic understanding of systems engineering, from a general applications perspective. The theory, methods, tools, techniques, and applications of systems engineering are covered in the book. The interactions between systems engineering and other disciplines are discussed. The book presents a comprehensive compilation of practical systems engineering models. The application and recognition of systems engineering continue to grow rapidly. In spite of this, there is limited collation of the various systems engineering models available. These models were developed explicitly for applications in various fields of endeavor. Notable among the models covered in this book are the V-model, the Tornado model, the Spiral model, the DEJI model, and the waterfall model. There are also many in-house and customized models developed by companies specifically for organizational needs. This book is intended to boost the collection of published titles on systems engineering. Although systems engineering models are used widely, there is a limited availability of titles dedicated specifically to how to use the models. A better understanding of the models, through a comprehensive reference book, will make systems engineering to be more visible, embraced, and applied across the spectrum. Not many books currently on the market address the availability and usability of systems engineering models as a tool for operational efficiency and organizational effectiveness. A system is defined as the collection of interrelated elements (subsystems) whose collective output is higher than the sum of the outputs of the individual elements. This aptly defines how everyday efforts manifest themselves. A sports team is a good example of this. The performance of a team is the result of the collective efforts of the players rather than the output of a single player. Without a structured guidance, we are likely to focus on the individual elements rather than the overall system. A key benefit for readers of this book is the ability to get better results due to systems efficiency and effectiveness. A system, in this regard, facilitates a full consideration and appreciation of all the obvious and subtle elements of any process. Some of the models espoused in this book include the V-model, the waterfall model, and the DEJI model (see www.DEJImodel.com for model details). Based on the systems integration component of the DEJI model, it is the featured model in this book as it cuts across applications of the other models.

# *Acknowledgments*

My immense thanks go to all my academic colleagues and professional mentors, who, over the years, have introduced me to the fine art of systems thinking. My appreciation also goes to my students, who, for many decades, have held me accountable for a real-life practice of the systems view that I always profess in the classroom. Special thanks go to Jinan Andrews, Luke Farrell, Jay McDaniel, Michael Avery, Kirk Weigand, Katie MacGregor, and Lukas Cowen for their intellectual contributions that I harvested to buttress the contents and quality of this manuscript. I also thank my home front (my family) for their continuing accommodation of my incessant forays into new writing ventures.

# Author

**Adedeji B. Badiru** is professor of Systems Engineering at the Air Force Institute of Technology (AFIT). He was previously professor and Head of Systems Engineering and Management at AFIT, professor and Department Head of Industrial & Information Engineering at the University of Tennessee in Knoxville, and professor of Industrial Engineering and Dean of University College at the University of Oklahoma, Norman. He is a registered professional engineer (PE), a certified Project Management Professional (PMP), a fellow of the Institute of Industrial Engineers, and a Fellow of the Nigerian Academy of Engineering. He holds BS in Industrial Engineering, MS in Mathematics, and MS in Industrial Engineering from Tennessee Technological University, and PhD in Industrial Engineering from the University of Central Florida. He is a prolific author and has won national and international awards for his scholarly, leadership, and professional accomplishments. He is a member of several professional associations and scholastic honor societies.

# chapter one

# Systems introduction

## Introduction

A system is represented as consisting of multiple parts, all working together for a common purpose or goal. Systems can be small or large, simple or complex. Small devices can also be considered systems. Systems have inputs, processes, and outputs. Systems are usually explained using a model for a visual clarification inputs, process, and outputs. A model helps illustrate the major elements and their relationships. Figure 1.1 illustrates the basic model structure of a system.

Systems engineering is the application of engineering tools and techniques to the solutions of multifaceted problems through a systematic collection and integration of parts of the problem with respect to the life cycle of the problem. It is the branch of engineering concerned with the development, implementation, and use of large or complex systems. It focuses on specific goals of a system considering the specifications, prevailing constraints, expected services, possible behaviors, and structure of the system. It also involves a consideration of the activities required to ensure that the system's performance matches specified goals. Systems

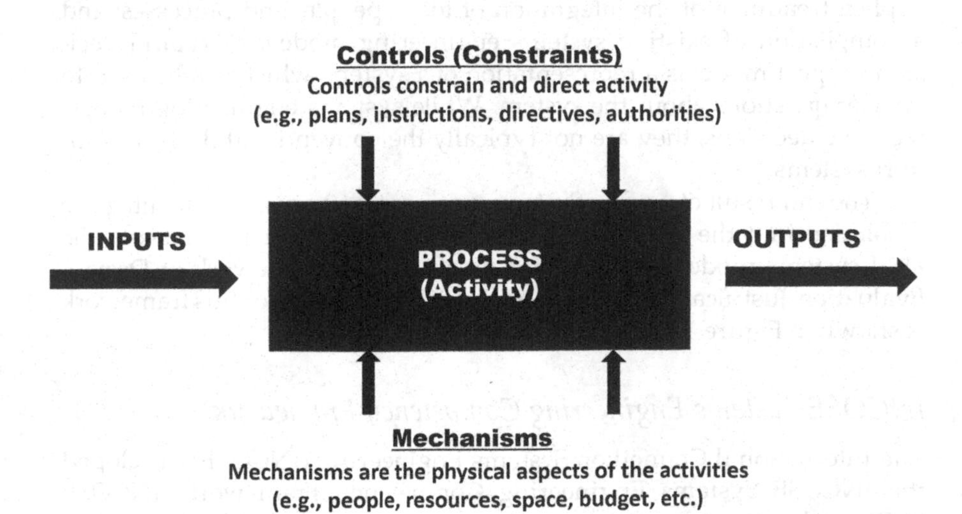


*Figure 1.1* Systems input, process, and output structure.

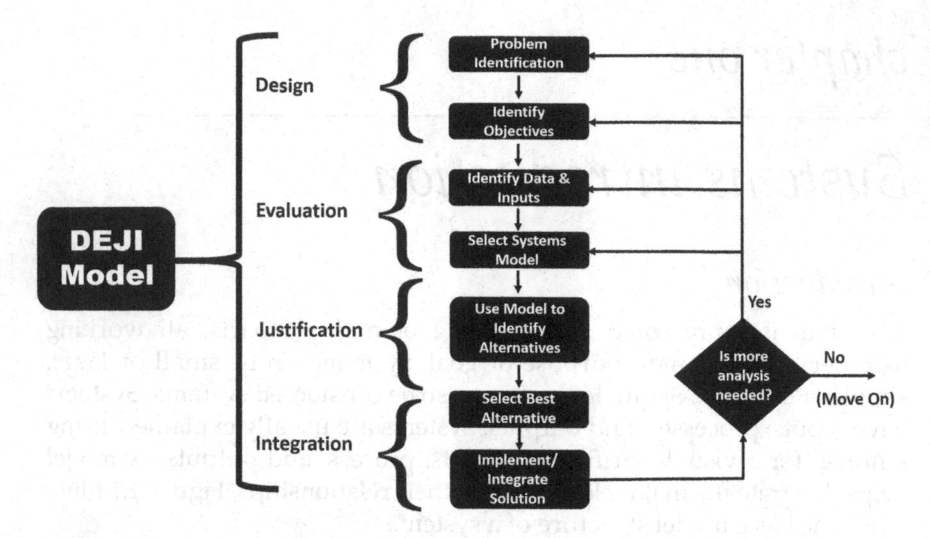


*Figure 1.2* Flowchart of DEJI model framework.

engineering addresses the integration of tools, people, and processes required to achieve a cost-effective and timely operation of the system. Some of the features of this book include solution to multifaceted problems; a holistic view of a problem domain; applications to both small and large problems; decomposition of complex problems into smaller manageable chunks; direct considerations for the pertinent constraints that exist in the problem domain; systematic linking of inputs to goals and outputs; explicit treatment of the integration of tools, people, and processes; and a compilation of existing systems engineering models. A typical decision support model is a representation of a system, which can be used to answer questions about the system. While systems engineering models facilitate decisions, they are not typically the conventional decision support systems.

The end result of using a Systems engineering approach is to integrate a solution into the normal organizational process. For that reason, the DEJI systems model is desired for its structured framework of Design, Evaluation, Justification, and Integration. The flowchart for this framework is shown in Figure 1.2.

## INCOSE Systems Engineering Competency Framework

The International Council on Systems Engineering (INCOSE) developed the INCOSE Systems Engineering Competency Framework (INCOSE SECF), which represents a worldview of five competency groupings with 36 competencies central to the profession of systems engineering. This includes evidence-based indicators of knowledge, skills, abilities, and behaviors across five levels of proficiency. INCOSE SECF supports a wide variety of usage scenarios including individual and organizational capability assessments. It enables organizations to tailor and derive their own competency models that address their unique challenges. The five competency groupings and their respective competencies are provided below:

- **Systems Engineering Management**
  1. Information Management
  2. Planning
  3. Monitoring and Control
  4. Risk and Opportunity Management
  5. Business and Enterprise Integration
  6. Decision Management
  7. Concurrent Engineering
  8. Configuration Management
  9. Acquisition and Supply
- **Professional**
  1. Communications
  2. Facilitation
  3. Ethics and Professionalism
  4. Coaching and Mentoring
  5. Technical Leadership
  6. Emotional Intelligence
  7. Negotiation
  8. Team Dynamics
- **Core Systems Engineering Principles**
  1. General Engineering
  2. Systems Thinking
  3. Capability Engineering
  4. Critical Thinking
  5. Systems Modeling and Analysis
  6. Life cycles
- **Integrating**
  1. Quality
  2. Finance
  3. Project Management
  4. Logistics
- **Technical**
  1. Requirements Definition
  2. Systems Architecting

3. Transition
4. Operations and Support
5. Design (for specific needs)
6. Integration
7. Interfaces
8. Verification
9. Validation

Each competency in the SECF framework has five proficiency levels as summarized below.

## *Awareness proficiency*

The person displays knowledge of key ideas associated with the competency area and understands key issues and their implications.

## *Supervised practitioner proficiency*

The person displays an understanding of the competency area and has some limited experience.

## *Practitioner*

The person displays both knowledge and practical experience of the competency area and can function without supervision on a day-to-day basis.

## *Lead practitioner*

The person displays extensive and substantial practical knowledge and experience of the competency area and provides guidance to others including practitioners encountering unusual situations.

## *Expert*

In addition to extensive and substantial practical experience and applied knowledge of the competency area, this individual contributes to and is recognized beyond the organizational or business boundary.

It should be noted that in many professional environments, it is believed that it takes about 15 years of practical experience to become an expert in any particular professional pursuit.

## *Systems attributes, factors, and indicators*

In any systems approach, the systems analyst must be cognizant of the attributes, factors, and indicators that fully describe the overall system.

For the technical competency part of SERC, the attributes, factors, and indicators of competency (knowledge and experience) are summarized below.

**Competency area**: Technical (requirements definition).

**Description**: To analyze the stakeholder needs and expectations to establish the requirements for a system.

**Purpose**: The requirements of a system describe the problem to be solved (its purpose; how it performs; how it is to be used, maintained, and disposed of; and what the expectations of the stakeholders are).

**For awareness**

- Describes different types of requirements (e.g., functional, nonfunctional, business).
- Explains why there is a need for good quality requirements.
- Identifies major stakeholders and their needs.
- Explains why managing requirements throughout the life cycle is important.

**Supervised practitioner**

- Identifies all stakeholders and their sphere of influence.
- Assists with the elicitation of requirements from stakeholders.
- Describes the characteristics of good quality requirements and provides examples.
- Describes different mechanisms used to gather requirements.

**Practitioner**

- Defines governing requirements elicitation and management plans, processes and appropriate tools and uses these to control and monitor requirements elicitation and management activities.
- Elicits and validates stakeholder requirements.
- Writes good quality, consistent requirements.

**Lead practitioner**

- Recognized, within the enterprise, as an authority in requirements elicitation and management techniques, contributing to best practice.
- Defines and documents enterprise-level policies, procedures, guidance, and best practice for requirements elicitation and management, including associated tools.
- Challenges appropriateness of requirements in a rational way.

**Expert**

- Recognized, beyond the enterprise boundary, as an authority in requirements elicitation and management techniques.

- Contributes to requirements elicitation and management best practice.
- Champions the introduction of novel techniques and ideas in requirements elicitation and management, producing measurable improvements.

**Competency area**: Systems engineering management (risk and opportunity management).

**Practitioner**

- Defines governing risk and opportunity management plans, processes and appropriate tools and uses these to control and monitor risk and opportunity management activities.
- Establishes a project risk and opportunity profile including context, probability, consequences, thresholds, priority, and risk action and status.
- Identifies, assesses, analyzes, and treats risks and opportunities for likelihood and consequence in order to determine magnitude and priority for treatment.
- Treats risks and opportunities effectively, considering alternative treatments and generating a plan of action when thresholds exceeds certain levels.
- Guides supervised practitioners in systems engineering risk and opportunity management.

**Lead practitioner**

- Recognized, within the enterprise, as an authority in systems engineering risk and opportunity management, contributing to best practice.
- Reviews and judges the tailoring of enterprise-level risk and opportunity management processes and associated work products to meet the needs of a project.
- Coordinates systems engineering risk and opportunity management across multiple diverse projects or across a complex system, with proven success.
- Establishes an enterprise risk profile including context, probability, consequences, thresholds, priority, and risk action and status.
- Coaches new and experienced practitioners in systems engineering risk and opportunity management.

For an effective application of SECF, all indicators must have the following properties:

- Start with action verbs.
- Are evidence based.

- Show progressions from lower to higher levels of proficiency.
- Can be mapped to a combination of knowledge, skills, abilities, behaviors, and experiences.
- Enable individuals to self-assess and acquire higher levels of proficiency.

## *The scope of systems engineering*

The synergy between software engineering and systems engineering is made evident by the integration of the methods and processes developed by one discipline into the culture of the other. Researchers from software engineering (Boehm, 1994) and systems engineering (Boehm 1981, Boehm et al., 2000, Boehm, Valerdi, and Honour, 2008, Boehm et al., 2005, Valerdi, 2005, Valerdi, 2014) have extensively promoted the integration of both disciplines but have faced roadblocks that result from the fundamental difference between the two (Pandikow and Törne, 2001).

However, the development of systems engineering standards has helped the crystallization of the discipline as well as the development of COSYSMO. Table 1.1 includes a list of the standards most influential to this effort.

The first U.S. military standard focused on systems engineering provided by the first definition of the scope of engineering management (MIL-STD-499A, 1969). It was followed by another military standard that provided guidance on the process of writing system specifications for military systems (MIL-STD-490A, 1985). These standards were influential in defining the scope of systems engineering in their time. Years later, the standard ANSI/EIA-632 *Processes for Engineering a System* (ANSI/EIA, 1999) provided a typical Systems engineering Work Breakdown Structure (WBS). The ANSI/EIA-632 standard was developed between 1994 and 1998 by a working group of industry associations, the INCOSE, and the U.S. Department of Defense with the intent to provide a standard for use by commercial enterprises, as well as government agencies and their contractors. It was designed to have a broader scope but less detail than

***Table 1.1*** Notable systems engineering standards

| Standard (year) | Title |
|---|---|
| MIL-STD-499A (1969) | *Engineering Management* |
| MIL-STD-490A (1985) | *Specification Practices* |
| ANSI/EIA-632 (1999) | *Processes for Engineering a System* |
| CMMI (2002) | *Capability Maturity Model Integration* |
| ANSI/EIA-731.1 (2002) | *Systems Engineering Capability Model* |
| ISO/IEC 15288 (2002) | *Systems Engineering – System Life Cycle Processes* |

previous systems engineering standards. Such lists provide, in much finer detail, the common activities that are likely to be performed by systems engineers in those organizations, but are generally not applicable outside of the companies in which they are created. In addition to organizational applicability, there are significant differences in different application domains, especially in space systems engineering (Valerdi, Wheaton, and Fortune, 2007).

The ANSI/EIA-632 standard provides the *what-if* systems engineering through five fundamental processes: (1) acquisition and supply, (2) technical management, (3) system design, (4) product realization, and (5) technical evaluation. These processes are the basis of the systems engineering effort profile developed for COSYSMO. The five fundamental processes are divided into 13 high-level process categories and further decomposed into 33 activities shown in Table 1.2.

This standard provides a generic list of activities which are generally applicable to individual companies, but each project should compare their own systems engineering WBS to the ones provided in the ANSI/EIA-632 standard to identify similarities and differences.

After defining the possible systems engineering activities used in COSYSMO, a definition of the system life cycle phases is needed to help bound the model and the estimates it produces. Because of the focus on systems engineering, COSYSMO employs some of the life cycle phases from ISO/IEC 15288 *Systems Engineering – System Life Cycle Processes* (ISO/IEC, 2002). These phases were slightly modified to reflect the influence of the aforementioned model, ANSI/EIA-632, and are shown in Figure 1.3. These life cycle phases help answer the *when* of systems engineering and COSYSMO. Understanding when systems engineering is performed relative to the system life cycle helps define anchor points for the model.

Life cycle models vary according to the nature, purpose, use, and prevailing circumstances of the product under development. Despite an infinite variety in system life cycle models, there is an essential set of characteristic life cycle phases that exists for use in the systems engineering domain. For example, the *Conceptualize* phase focuses on identifying stakeholder needs, exploring different solution concepts, and proposing candidate solutions. The *Development* phase involves refining the system requirements, creating a solution description, and building a system. The *Operational Test and Evaluation* phase involves verifying/validating the system and performing the appropriate inspections before it is delivered to the user. The *Transition to Operation* phase involves the transition to utilization of the system to satisfy the users' needs via training or handoffs. These four life cycle phases are within the scope of COSYSMO. The final two were included in the data collection effort but did not yield enough

***Table 1.2*** ANSI/EIA (1999) systems engineering processes and activities

| Fundamental processes | Process categories | Activities |
|---|---|---|
| Acquisition and supply | Supply process | (1) Product supply |
| | Acquisition process | (2) Product acquisition, (3) Supplier performance |
| Technical management | Planning process | (4) Process implementation strategy, (5) Technical effort definition, (6) Schedule and organization, (7) Technical plans, (8) Work directives |
| | Assessment process | (9) Progress against plans and schedules, (10) Progress against requirements, (11) Technical reviews |
| | Control process | (12) Outcomes management, (13) Information dissemination |
| System design | Requirements definition process | (14) Acquirer requirements, (15) Other stakeholder requirements, (16) System technical requirements |
| | Solution definition process | (17) Logical solution representations, (18) Physical solution representations, (19) Specified requirements |
| Product realization | Implementation process | (20) Implementation |
| | Transition to use process | (21) Transition to use |
| Technical evaluation | Systems analysis process | (22) Effectiveness analysis, (23) Trade-off analysis, (24) Risk analysis |
| | Requirements validation process | (25) Requirement statements validation, (26) Acquirer requirements, (27) Other stakeholder requirements, (28) System technical requirements, (29) Logical solution representations |
| | System verification process | (30) Design solution verification, (31) End product verification, (32) Enabling product readiness |
| | End products validation process | (33) End products validation |

data to perform a calibration. These phases are *Operate, Maintain, or Enhance,* which involves the actual operation and maintenance of the system required to sustain system capability, and *Replace or Dismantle,* which involves the retirement, storage, or disposal of the system.

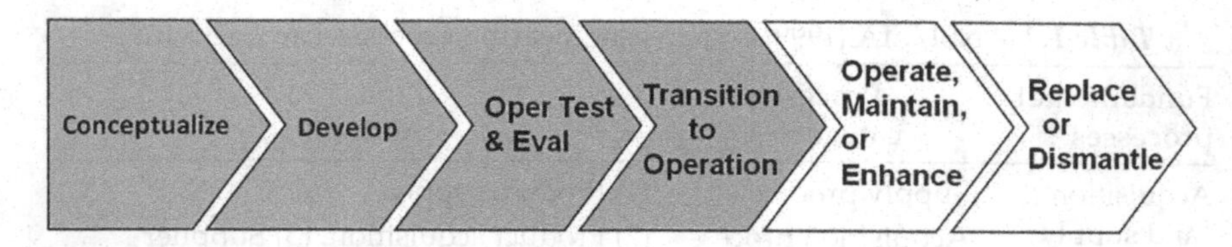


*Figure 1.3* COSYSMO system life cycle phases.

## Systems definitions and attributes

One definition of systems project management offered here is stated as follows:

> Systems project management is the process of using systems approach to manage, allocate, and time resources to achieve systems-wide goals in an efficient and expeditious manner.

The above definition calls for a systematic integration of technology, human resources, and work process design to achieve goals and objectives. There should be a balance in the synergistic integration of humans and technology. There should not be an overreliance on technology, nor should there be an overdependence on human processes. Similarly, there should not be too much emphasis on analytical models to the detriment of common-sense human-based decisions.

Systems engineering is growing in appeal as an avenue to achieve organizational goals and improve operational effectiveness and efficiency. Researchers and practitioners in business, industry, and government are

> Systems engineering is the application of engineering to solutions of a multifaceted problem through a systematic collection and integration of parts of the problem with respect to the life cycle of the problem. It is the branch of engineering concerned with the development, implementation, and use of large or complex systems. It focuses on specific goals of a system considering the specifications, prevailing constraints, expected services, possible behaviors, and structure of the system. It also involves a consideration of the activities required to assure that the system's performance matches the stated goals. Systems engineering addresses the integration of tools, people, and processes required to achieve a cost-effective and timely operation of the system.

all clamoring collaboratively for systems engineering implementations. So, what is systems engineering? Several definitions exist. Below is one quite comprehensive definition:

Logistics can be defined as the planning and implementation of a complex task; the planning and control of the flow of goods and materials through an organization or manufacturing process; or the planning and organization of the movement of personnel, equipment, and supplies. Complex projects represent a hierarchical system of operations. Thus, we can view a project system as collection of interrelated projects all serving a common end goal. Consequently, we present the following universal definition:

> Project systems logistics is the planning, implementation, movement, scheduling, and control of people, equipment, goods, materials, and supplies across the interfacing boundaries of several related projects.

Conventional project management must be modified and expanded to address the unique logistics of project systems.

## Systems constraints

Systems management is the pursuit of organizational goals within the constraints of time, cost, and quality expectations. The iron triangle model shows that project accomplishments are constrained by the boundaries of quality, time, and cost. In this case, quality represents the composite collection of project requirements. In a situation where precise optimization is not possible, there will have to be trade-offs between these three factors of success. The concept of iron triangle is that a rigid triangle of constraints encases the project. Everything must be accomplished within the boundaries of time, cost, and quality. If better quality is expected, a compromise along the axes of time and cost must be executed, thereby altering the shape of the triangle. The trade-off relationships are not linear and must be visualized in a multidimensional context. Scope requirements determine the project boundary, and trade-offs must be done within that boundary. If we label the eight corners of the box as (a), (b), (c), …, (h), we can iteratively assess the best operating point for the project. For example, we can address the following two operational questions:

1. From the point of view of the project sponsor, which corner is the most desired operating point in terms of combination of requirements, time, and cost?

2. From the point of view of the project executor, which corner is the most desired operating point in terms of combination of requirements, time, and cost?

Note that all the corners represent extreme operating points. We notice that point (e) is the do-nothing state, where there are no requirements, no time allocation, and no cost incurrence. This cannot be the desired operating state of any organization that seeks to remain productive. Point (a) represents an extreme case of meeting all requirements with no investment of time or cost allocation. This is an unrealistic extreme in any practical environment. It represents a case of getting something for nothing. Yet, it is the most desired operating point for the project sponsor. By comparison, point (c) provides the maximum possible for requirements, cost, and time. In other words, the highest levels of requirements can be met if the maximum possible time is allowed and the highest possible budget is allocated. This is an unrealistic expectation in any resource-conscious organization. You cannot get everything you ask for to execute a project. Yet, it is the most desired operating point for the project executor. Considering the two extreme points of (a) and (c), it is obvious that the project must be executed within some compromise region within the scope boundary. A graphical analysis can reveal a possible view of a compromise surface with peaks and valleys representing give-and-take trade-off points within the constrained box. The challenge is to come up with some analytical modeling technique to guide decision-making over the compromise region. If we could collect sets of data over several repetitions of identical projects, then we could model a decision surface that can guide future executions of similar projects. Such typical repetitions of an identical project are most readily apparent in construction projects, for example residential home development projects.

Systems influence philosophy suggests the realization that you control the internal environment while only influencing the external environment. The inside (controllable) environment is represented as a black box in the typical input–process–output relationship. The outside (uncontrollable) environment is bounded by a cloud representation. In the comprehensive systems structure, inputs come from the global environment, are moderated by the immediate outside environment, and are delivered to the inside environment. In an unstructured inside environment, functions occur as blobs. A "blobby" environment is characterized by intractable activities where everyone is busy, but without a cohesive structure of input–output relationships. In such a case, the following disadvantages may be present:

- Lack of traceability
- Lack of process control

- Higher operating cost
- Inefficient personnel interfaces
- Unrealized technology potentials

Organizations often inadvertently fall into the blobs structure because it is simple, low-cost, and less time-consuming, until a problem develops. A desired alternative is to model the project system using a systems value-stream structure. This uses a proactive and problem-preempting approach to execute projects. This alternative has the following advantages:

- Problem diagnosis is easier.
- Accountability is higher.
- Operating waste is minimized.
- Conflict resolution is faster.
- Value points are traceable.

## Systems value modeling

A technique that can be used to assess overall value-added components of a process improvement program is the systems value model (SVM). The model provides an analytical decision aid for comparing process alternatives. Value is represented as a $p$-dimensional vector:

$$V = f\left(A_1, A_2, \ldots, A_p\right)$$

where $A = \left(A_1, \ldots, A_n\right)$ is a vector of quantitative measures of tangible and intangible attributes. Examples of process attributes are quality, throughput, capability, productivity, cost, and schedule. Attributes are considered to be a combined function of factors, $x_1$, expressed as

$$A_k\left(x_1, x_2, \ldots, x_{m_k}\right) = \sum_{i=1}^{m_k} f_i\left(x_i\right)$$

where $\{x_i\}$ = set of $m$ factors associated with attribute $A_k\left(k = 1, 2, \ldots, p\right)$ and $f_i$ = contribution function of factor $x_i$ to attribute $A_k$. Examples of factors include reliability, flexibility, user acceptance, capacity utilization, safety, and design functionality. Factors are themselves considered to be composed of indicators, $v_i$, expressed as

$$x_i\left(v_1, v_2, \ldots, v_n\right) = \sum_{j=1}^{n} z_i\left(v_i\right)$$

where $\{v_j\}$ = set of $n$ indicators associated with factor $x_i\,(i=1,2,\ldots,m)$ and $z_j$ = scaling function for each indicator variable $v_j$. Examples of indicators are project responsiveness, lead time, learning curve, and work rejects. By combining the above definitions, a composite measure of the value of a process can be modeled as

$$V = f\left(A_1, A_2, \ldots, A_p\right)$$

$$= f\left\{\left[\sum_{i=1}^{m_1} f_i\left(\sum_{j=1}^{n} z_j\left(v_j\right)\right)\right]_1, \left[\sum_{i=1}^{m_2} f_i\left(\sum_{j=1}^{n} z_j\left(v_j\right)\right)\right]_2, \ldots, \left[\sum_{i=1}^{m_k} f_i\left(\sum_{j=1}^{n} z_j\left(v_j\right)\right)\right]_p\right\}$$

where $m$ and $n$ may assume different values for each attribute. A subjective measure to indicate the utility of the decision maker may be included in the model by using an attribute weighting factor, $w_i$, to obtain a weighted $PV$:

$$PV_w = f\left(w_1 A_1, w_2 A_2, \ldots, w_p A_p\right)$$

where

$$\sum_{k=1}^{p} w_k = 1, \quad (0 \le w_k \le 1).$$

With this modeling approach, a set of process options can be compared on the basis of a set of attributes and factors.

To illustrate the model above, suppose three IT options are to be evaluated based on four attribute elements: *capability, suitability, performance,* and *productivity.* For this example, based on the equations, the value vector is defined as

$$V = f\left(\text{capability}, \text{suitability}, \text{performance}, \text{productivity}\right)$$

*Capability:* The term *"capability"* refers to the ability of IT equipment to satisfy multiple requirements. For example, a certain piece of IT equipment may only provide computational service. A different piece of equipment may be capable of generating reports in addition to computational analysis, thus increasing the service variety that can be obtained. In the analysis, the levels of increase in service variety from the three competing equipment types are 38%, 40%, and 33%, respectively.

*Suitability:* This refers to the appropriateness of the IT equipment for current operations. For example, the respective percentages of operating scope for which the three options are suitable for are 12%, 30%, and 53%, respectively.

*Performance:* In this context, this refers to the ability of the IT equipment to satisfy schedule and cost requirements. In the example, the three options can, respectively, satisfy requirements on 18%, 28%, and 52% of the typical set of jobs.

*Productivity:* This can be measured by an assessment of the performance of the proposed IT equipment to meet workload requirements in relation to the existing equipment. For example, the three options, respectively, show normalized increases of 0.02, –1.0, and –1.1 on a uniform scale of productivity measurement. Option C is the best "value" alternative in terms of suitability and performance. Option B shows the best capability measure, but its productivity is too low to justify the needed investment. Option A offers the best productivity, but its suitability measure is low. The analytical process can incorporate a lower control limit into the quantitative assessment such that any option providing value below that point will not be acceptable. Similarly, a minimum value target can be incorporated into the graphical plot such that each option is expected to exceed the target point on the value scale.

The relative weights used in many justification methodologies are based on subjective propositions of decision makers. Some of those subjective weights can be enhanced by the incorporation of utility models. For example, the weights could be obtained from utility functions. There is a risk of spending too much time maximizing inputs at "point-of-sale" levels with little time defining and refining outputs at the "wholesale" systems level.

A Systems view of operations is essential for every organization. Without a Systems view, we cannot be sure we are pursuing the right outputs. A systems approach allows for a multidimensional analysis of any endeavor, considering many of the typical "ilities" of systems engineering as listed below:

- Affordability
- Practicality
- Desirability
- Configurability
- Modularity
- Reliability
- Desirability
- Maintainability
- Testability

- Transmittability
- Reachability
- Quality
- Agility

A systems engineering plan is essential for the following reasons:

1. Description of the system being developed
2. Description of team structure and responsibilities
3. Identification of all project stakeholders
4. Description of tailored technical activities in each phase
5. Documentation of decisions and technical implementation
6. Establishment of technical metrics and measurements (Who, What, When, Where, Which, How, Why)

Now that we have explained some of the characteristics of a system, we can move on to specific applications and other considerations in the chapters that follow.

## *References*

ANSI/EIA-632-1988. (1999). *Processes for Engineering a System*. American National Standards Institute, New York.

ANSI/EIA-731.1. (2002). *Systems Engineering Capability Model*. American National Standards Institute, New York.

Boehm, B. W. (1981). *Software Engineering Economics*. Prentice Hall, Upper Saddle River, NJ.

Boehm, B. W. (1994). Integrating software engineering and systems engineering, *The Journal of NCOSE* 1(1), 147–151.

Boehm, B. W., Abts, C., Brown, A. W., Chulani, S., Clark, B., Horowitz, E., Madachy, R., Reifer, D. J., and Steece, B. (2000). *Software Cost Estimation with COCOMO II*. Prentice Hall, Upper Saddle River, NJ.

Boehm, B. W., Valerdi, R., and Honour, E. (2008). The ROI of systems engineering: Some quantitative results for software-intensive systems, *Systems Engineering* 11(3), 221–234.

Boehm, B. W., Valerdi, R., Lane, J., and Brown, A. W. (2005). COCOMO suite methodology and evolution, *CrossTalk - The Journal of Defense Software Engineering* 18(4), 20–25.

CMMI. (2002). *Capability Maturity Model Integration - CMMI-SE/SW/IPPD/SS, V1.1*. Carnegie Mellon - Software Engineering Institute, Pittsburg, PA.

ISO/IEC 15288. (2002). *Systems Engineering – System Life Cycle Processes*, First Edition, U.S. Department of Defense, Washington, DC.

MIL-STD-499A. (1969). *Engineering Management*. U.S. Department of Defense, Washington, DC.

MIL-STD-490A. (1985). *Specification Practices*. U.S. Department of Defense, Washington, DC.

Pandikow, A. and Törne, A. (2001). Integrating modern software engineering and systems engineering specification techniques. *14th International Conference on Software & Systems Engineering and Their Applications*, Vol. 2, Cocoa Beach, FL.
Valerdi, R. (2005). The constructive systems engineering cost estimation model (COSYSMO). *Ph. D. Dissertation*, University of Southern California, California.
Valerdi, R. (2014). Systems engineering cost estimation with a parametric model. In A. B. Badiru (Ed.), *Handbook of Industrial and Systems Engineering*. CRC Press, Boca Raton, FL.
Valerdi, R., Wheaton, M. J., and Fortune, J. (2007). Systems engineering cost estimation for space systems. *AIAA Space, AIAA 2007-6001*, Los Angeles, CA.

# chapter two

# Model-based systems engineering

## Introduction

A good model is a good guide for process execution. Model-Based Engineering (MBE) is a general approach to engineering that uses models as an integral part of the technical baseline that includes the requirements, analysis, design, implementation, and verification of a capability, system, and/or product throughout a systems life cycle. Model-Based Systems Engineering (MBSE) focuses MBE specifically on the approach of systems engineering. MBSE is the formalized application of modeling to support system requirements, design, analysis, verification, and validation activities beginning in the conceptual design phase and continuing throughout development and later life cycle phases. MBSE provides significant opportunities for improved productivity, efficiency, effectiveness, and product quality. This is a relatively new approach to the application of systems engineering and has received growing attention and acceptance.

## What is a model?

A model facilitates repeatability and consistency of activities in a systems environment. In general terms, a model is a simplified version of a concept, phenomenon, relationship, structure, organization, enterprise, or system. It can be a graphical, mathematical, flow diagram, or physical representation. A model is an abstraction of reality by focusing on the necessary components and eliminating or minimizing unnecessary components. The objectives of a representative model include the following:

- Facilitate understanding.
- Aid in decision-making.
- Examine "what if" scenarios in a decision environment.
- Explain the characteristics of a system.
- Control events.
- Predict events.
- Describe systems profile.
- Prescribe actions in the system.
- Diagnose problems in a system.

## *Elements of MBSE*

- Provides a repeatable template for systems actions
- Formalizes the practice of systems development through the use of models
- Broad scope with respect to multiple modeling domains across the life cycle
- Guides both horizontal and vertical integration
- Facilitates improvements in quality and productivity
- Lowers the risk of operations
- Inherently embeds rigor and precision
- Enforces communication among internal and external teams and the customer
- Enhances the management of complexity

The replication of models horizontally and vertically provides a comprehensive view of the systems environment, as illustrated in Figure 2.1. In the figure, the vertical flow goes from the component level to the system level and on to the operational level. The horizontal integration flows through the stages of 1, 2, 3, and 4. The MBSE interfaces and inner workings of are represented by the geared systems engineering elements. The overall illustration essentially represents a meta-model structure, whereby each inner element is, itself, a representative model

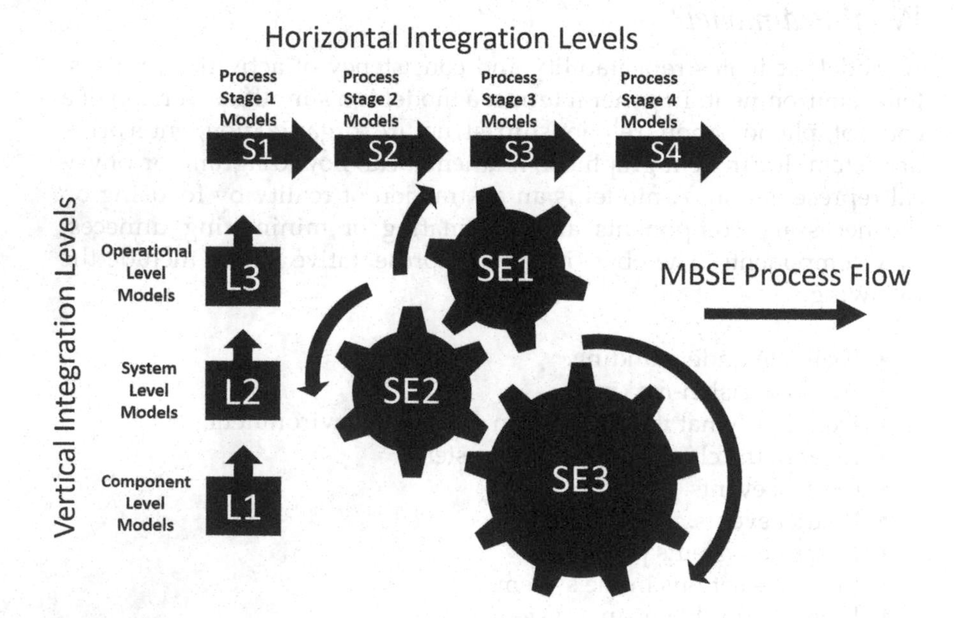


*Figure 2.1* MBSE vertical and horizontal integration.

of a function or process. For example, L2 may represent an organization chart of functions, while L3 may represent a flow diagram. Further, the stage S3 may be a quantitatively driven optimization model. The MBSE approach allows a hybrid use of qualitative and quantitative tools and techniques.

MBSE provides a means for driving more systems engineering depth without increasing costs. Data-centric specifications within MBSE enable automation and optimization, and unity of focus on value-added tasks. It also ensures a balanced approach to functions as well as an increased level of systems understanding. Systems understanding can be achieved through integrated analytics that are tied to a model-centric technical baseline. MBSE also drives a consistent specification across the design spectrum.

The key to a successful model-based approach is scoping the problem. Pertinent questions include:

What is expected out of the model?
What level of fidelity is desired to be accomplished?
What are the success criteria for the application of the model?

Scoping and managing a modeling effort is both an art and a science. Since driving change in an organization takes time, interpersonal relationship, and commitment to continuous investment, the mix of art and science is very important. Figure 2.2 gives an example of the application of the MBSE approach to the pursue of innovation in a defense acquisition program for the purpose of ensuring resilient position, navigation, and timing of components in the acquisition of new aircraft technology. For a continuity of the model, the sustainment end point will loop back to the beginning for the purpose of identifying new needs for the next cycle. Similarly, Figure 2.3 illustrates a semantic network of elements

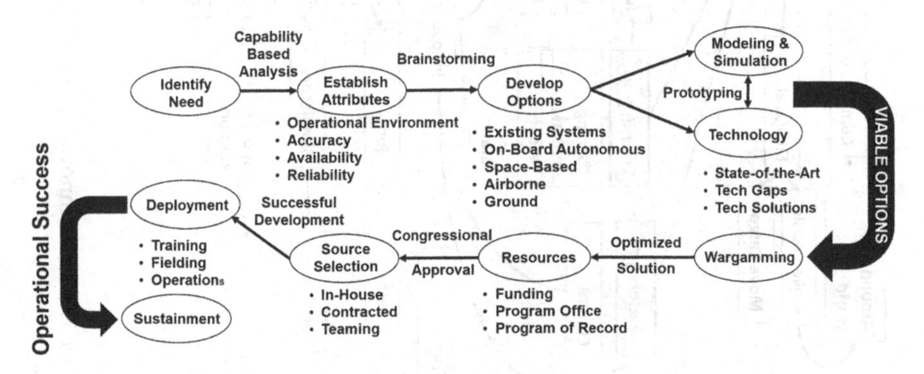


***Figure 2.2*** MBSE flow example for defense acquisition.

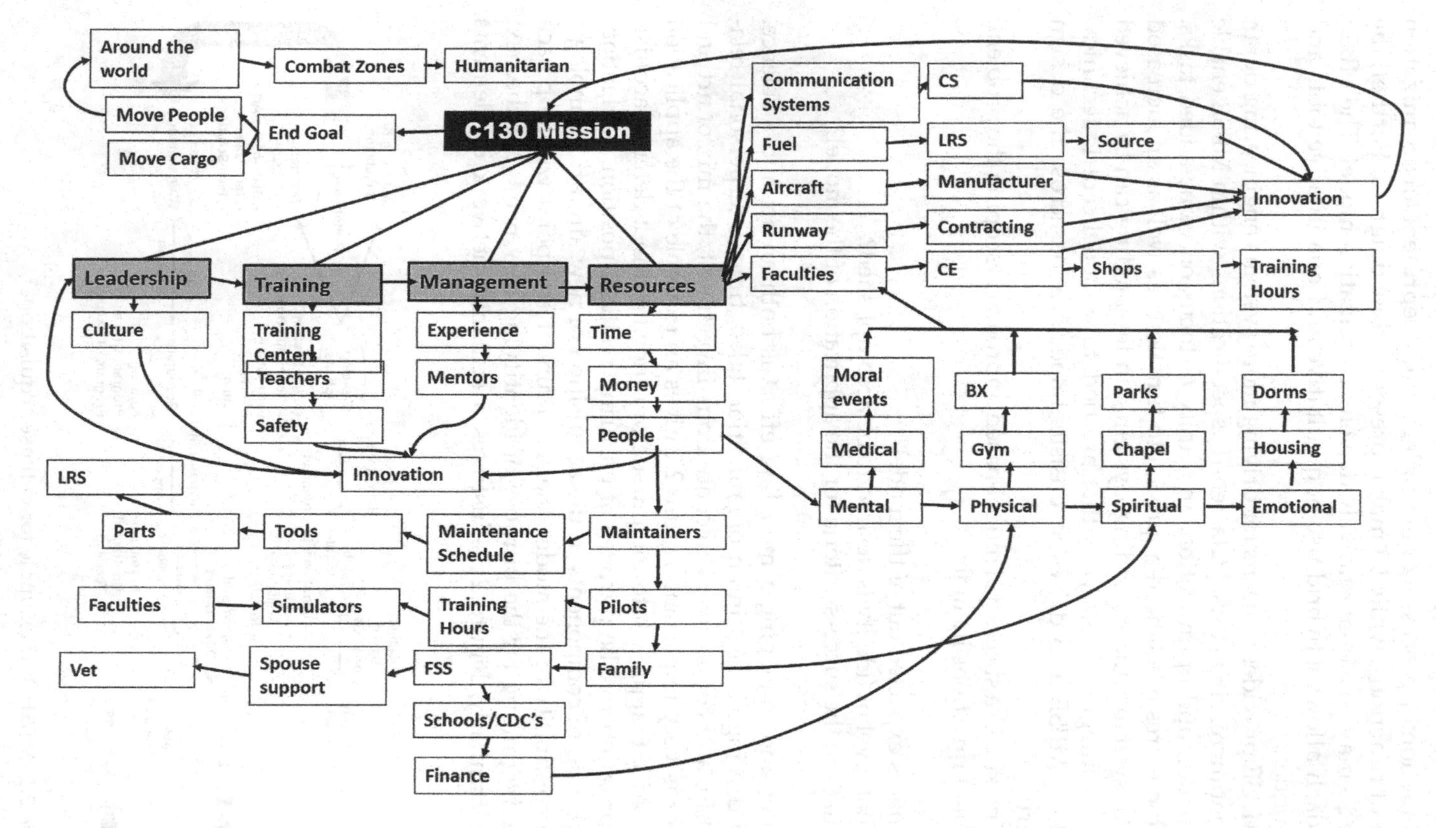


*Figure* 2.3 MBSE innovation flowchart for aircraft mission application.

for driving innovation in a C130 military mission. The characteristics of MBSE include:

- Set of interconnected models
- Models that are an abstraction of reality
- Vetted structure
- Behavior and requirements
- Standard language
- Graphical notation
- Syntax
- Semantics
- Visual focus
- Static and dynamic components
- Shared system information base

In moving from document-centric approach to model-centric approach, the linkages as articulated in Figure 2.4 are essential. Figure 2.5 illustrates an example of MBSE modeling of an automobile system. Table 2.1 shows a comparison of Document-Based and Model-Based Systems.

It shows a flow and interplay of various quantitative and qualitative aspects of operating the automobile system. The automobile image

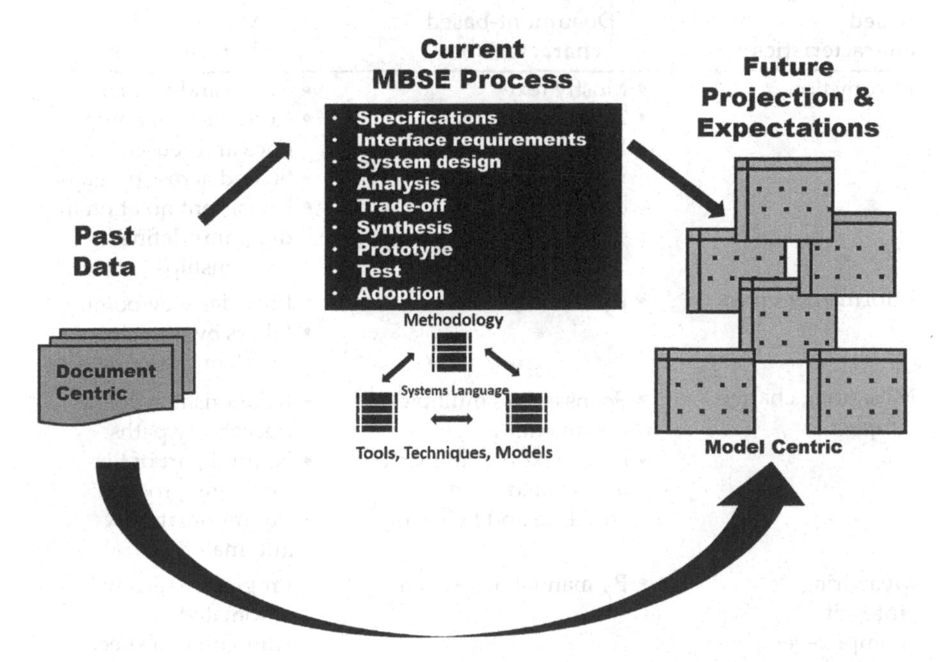


*Figure 2.4* MBSE data flow (past, present, future).

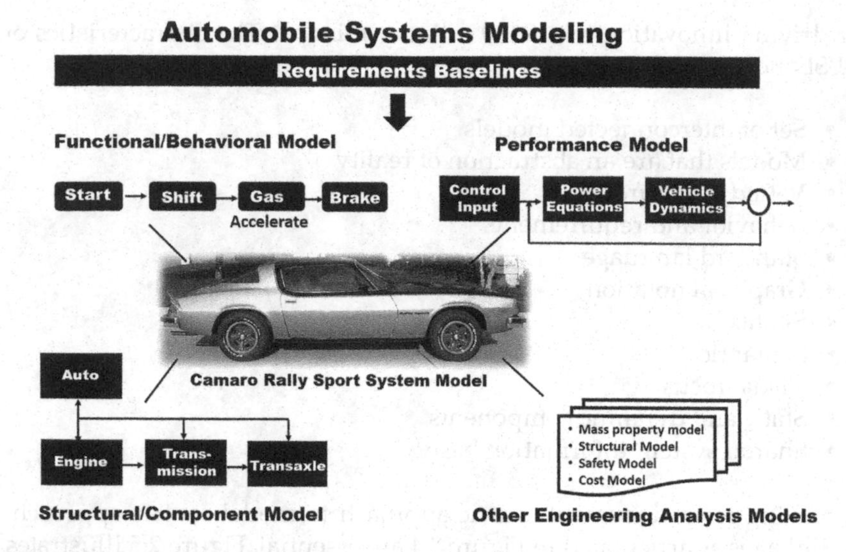


*Figure 2.5* MBSE automotive systems modeling.

*Table 2.1* Comparison of document- and model-based systems

| Information-based characteristics | Document-based characteristics | Model-based characteristics |
|---|---|---|
| Information | • Mostly text<br>• Add hoc diagrams<br>• Loosely coupled, repeated in multiple documents | • Visual and textual<br>• Constructs defined once and reused<br>• Shared across domains<br>• Consistent notation in diagrams defined relationships |
| Information views | • By document | • Provides viewpoints<br>• Filters by domain, problem space, etc. |
| Measuring change impact | • Spans across multiple documents<br>• Often text requirements are isolated from structure and behavior | • Relationships define traceability paths<br>• Natural part of the modeling process<br>• Programmatically automated |
| Measuring integrity: completeness, quality, and accuracy | • By manual inspection | • Programmatically automated<br>• Animation of spec |

used in the illustration is a 1976 General Motors Camaro Rally Sport once owned by the auto, with which numerous mechanical experimentations were conducted.

Systems modeling helps accomplish the following:

- Improved system and software
  - Specification
  - Visualization
  - Architecture
  - Construction
  - Simulation and test
  - Documentation
  - Validation and verification
- Improved communications
  - Enhanced knowledge capture and transfer
  - Training support
- Improved design quality
  - Decreased ambiguity
  - Increased precision
  - Supports evaluation of consistency, correctness, and completeness
  - Supports evaluation of trade space

Involvement of internal and external stakeholders is essential. The categories of stakeholders can be extensive, but basically include the following:

1. Customers
2. Project managers
3. Designers, developers, integrators
4. Vendors
5. Testers
6. Regulators

The impact of MBSE extends to several organizational aspects including:

- Systems architecture
- Cost
- Performance
- CAD (computer-aided design)
- Manufacturing
- Electronics
- Software
- Verification

With MBSE, decision makers will have more information and options from which to draw conclusions. Integrated analytics models will both increase the amount of information available to decision makes as well as help decision makers make sense of the information. Tools to explore, visualize, and understand a complex trade space, rooted in MBSE can provide early insight into the impact of decisions ranging from technical solutions to complex public policies.

The primary focus of many industry efforts is to move toward MBSE by integrating data through SE models. By bringing together varied but related models into a data rich, architecture centric environment, new levels of systems understanding can be achieved. MBSE forms a means to achieve integration.

## *Process improvement in model-based systems*

Model-Based Systems call for incorporating models into the overall operational scheme of the organization. Over the past few decades, a process approach has come to dominate our view of how to conceptualize and organize work, from a systems perspective (Heminger, 2014). Current approaches to management, such as Business Process Reengineering (BPR), Lean, and SixSigma are all based on this concept. Indeed, it seems almost axiomatic today to assume that this is the correct way to understand organizational work. Yet, each of these approaches seems to say different things about processes. What do they have in common that supports using a process approach? And, what do their different approaches tell us about different types of problems with the management of organizational work. To answer these questions, it may help to take a historical look at how work has been done since before the industrial revolution up to today.

Prior to the industrial revolution, work was done largely by craftsmen, who underwent a process of becoming skilled in their trade of satisfying customers wants and needs. Typically, they started as apprentices, where they learned the rudiments of their craft from beginning to end, moved on to become journeymen, then craftsmen as they become more knowledgeable, finally, reaching the pinnacle of their craft as master craftsmen. They grew both in knowledge of their craft and in understanding what their customers wanted. In such an arrangement, organizational complexity was low, with a few journeymen and apprentices working for a master craftsman. But, because work by craftsmen was slow and labor intensive, only a few of the very wealthiest people could have their needs for goods met. Most people did not have access to the goods that the few at the top of the economic ladder were able to get. There was a longstanding and persistent unmet demand for more goods.

This unmet demand, coupled with a growing technological capability, provided the foundations for the industrial revolution. Manufacturers developed what Smith (1776) called the "division of labor," in which complex tasks were broken down into simple tasks, automated where possible, and supervisors/managers were put in place to see that the pieces came together as a finished product. As we moved further into the industrial revolution, we continued to increase our productivity and the complexity of our factories. With the huge backlog of unmet demand, there was a willing customer for most of what was made. But, as we did this, an important change was taking place in how we made things. Instead of having a master craftsman in charge who knew both how to make goods as well as what the customers wanted and needed, we had factory supervisors, who learned how to make the various parts of the manufactured goods come together. Attention and focus began to turn inward from the customers to the process of monitoring and supervising complex factory work.

Over time, our factories became larger and ever more complex. More and more management attention needed to be focused inward on the issues of managing this complexity to turn out ever higher quantities of goods. In the early years of the twentieth century, Alfred Sloan, at General Motors, did for management what the industrial revolution had done for labor. He broke management down into small pieces, and assigned authority and responsibility tailored to those pieces. This allowed managers to focus on small segments of the larger organization, and to manage according to the authority and responsibility assigned. Through this method, General Motors was able to further advance productivity in the workplace. Drucker (1993) credits this internal focus on improved productivity for the creation of the middle class over the past 100 years. Again, because of the long-standing unmet demand, the operative concept was that if you could make it, you could sell it. The ability to turn out huge quantities of goods culminated in the vast quantities of goods created in the United States during and immediately following World War II. This was added to by manufacturers in other countries which came back on line after having their factories damaged or destroyed by the effects of the war. As they rebuilt and began producing again, they added to the total quantities of goods being produced.

Then, something happened that changed everything. Supply started to outstrip demand. It didn't happen everywhere evenly, either geographically, or by industry. But, in ever-increasing occurrences, factories found themselves supplying more than people were demanding. We had reached a tipping point. We went from a world where demand outpaced supply to a world where more and more supply outpaced demand (Hammer and Champy, 1993). Not everything being made was going to sell, at least not for a profit. When supply outstrips demand, customers can choose. And, when customers can choose, they will choose. Suddenly, manufacturers were faced with what Hammer and Champy call the "3 Cs": customers, competition, and change (Hammer and Champy, 1993). Customers were choosing among competing products, in a world of constant technological change. To remain in business, it was now necessary to produce those products that customers will choose. This required knowing what customers wanted. But, management and the structure of organizations from the beginning of the industrial revolution have been largely focused inward, on raising productivity and making more goods for sale. Managerial structure, information flows, and decision points were largely designed to support the efficient manufacturing of more goods, not on tailoring productivity to the needs of choosy customers.

### *Business process reengineering*

A concept was needed that would help organizations focus on their customers and their customers' needs. A process view of work provided a path for refocusing organizational efforts on meeting customer needs and expectations. On one level, a process is simply a series of steps, taken in some order, to achieve some result. Hammer and Champy, however, provided an important distinction in their definition of a process. They defined it as "a collection of activities that takes one or more inputs and creates an output that is of value to the customer" (1993). By adding the customer to the definition, Hammer and Champy provided a focus back on the customer, where it had been prior to the industrial revolution. In their 1993 book, *Reengineering the Corporation: A Manifesto for Business Revolution*, Hammer and Champy advocated BPR, which they defined as "the fundamental rethinking and radical redesign of business processes to achieve dramatic improvements in critical, contemporary measures of performance...." In that definition, they identified four words that they believed were critical to their understanding of reengineering. Those four words were "fundamental," "radical," "dramatic," and "processes." In the following editions of their book, which came out in 2001 and 2003, they revisited this definition and decided that the key word underlying all of their efforts was the word "process." And, with process defined as "taking inputs, and turning them into outputs of value to a customer," customers and what customers' value are the focus of their approach to reengineering.

Hammer and Champy viewed BPR as a means to rethink and redesign organizations to better satisfy their customers. BPR would entail challenging the assumption under which the organization had been operating, and to redesign around their core processes. They viewed the creative use of information technology as an enabler that would allow them to provide the information capabilities necessary to support their processes while minimizing their functional organizational structure.

## *Lean*

At roughly the same time that this was being written by Hammer and Champy, Toyota was experiencing increasing success and buyer satisfaction through its use of Lean, which is a process view of work focused on removing waste from the value stream. Womack and Jones (2003) identified the first of the Lean principles as value. And, they state, "Value can only be defined by the ultimate customer." So, once again, we see a management concept that leads organizations back to focus on their customers. Lean is all about identifying waste in a value stream (similar to Hammer and Champy's process) and removing that waste wherever possible. But, the identification of what is waste can only be determined by what contributes or doesn't contribute to value, and value can only be determined by the ultimate customer. So, once again, we have a management approach that refocuses organizational work on the customers and their values.

Lean focuses on five basic concepts: value, the value stream, flow, pull, and perfection. "Value," which is determined by the ultimate customer, and the "value stream" can be seen as similar to Hammer and Champy's "process," which focuses on adding value to its customers. "Flow" addresses the passage of items through the value stream, and it strives to maximize the flow of quality production. "Pull" is unique to Lean and is related to the "just-in-time" nature of current manufacturing. It strives to reduce in-process inventory that is often found in large manufacturing operations. "Perfection" is the goal that drives Lean. It is something to be sought after, but never to be achieved. Thus, perfection provides the impetus for constant process improvement.

## *Six Sigma*

In statistical modeling of manufacturing processes, sigma refers to the number of defects per given number of items created. Six Sigma refers to a statistical expectation of 3.4 defects per million items. General Electric adopted this concept in the development of the Six Sigma management strategy in 1986. While statistical process control can be at the heart of a Six Sigma program, General Electric and others have broadened its use to include other types of error reduction as well. In essence, Six Sigma is a program focused on reducing errors and defects in an organization. While Six Sigma does not explicitly refer back to the customer for its source of creating quality, it does address the concept of reducing errors and variations in specifications. Specifications can be seen as coming from customer requirements; so again, the customer becomes key to success in a Six Sigma environment.

Six Sigma makes the assertion that quality is achieved through continuous efforts to reduce variation in process outputs. It is based on collecting and analyzing data, rather than depending on hunches or guesses as a basis for making decisions. It uses the steps define, measure, analyze, improve, and control (DMAIC) to improve existing processes. To create new processes, it uses the steps define, measure, analyze, design, and verify (DMADV). Unique to this process improvement methodology, Six Sigma uses a series of karate-like levels (yellow belts, green belts, black belts, and master black belts) to rate practitioners of the concepts in organizations. Many companies who use Six Sigma have been satisfied by the improvements that they have achieved. To the extent that output variability is an issue for quality, it appears that Six Sigma can be a useful path for improving quality.

## *Selecting a methodology*

From the above descriptions, it is clear that while each of these approaches uses a process perspective, they address different problem sets, and they suggest different remedies. BPR addresses the problem of getting a good process for the task at hand (Table 2.2). It recognizes that many business processes over the years have been designed with an internal focus, and it uses a focus on the customer as a basis for redesigning processes that explicitly address what customers need and care about. This approach would make sense where organizational processes have become focused on internal management needs, or some other issues, rather than on the needs of the customer.

The Lean methodology came out of the automotive world, and is focused on gaining efficiencies in manufacturing. Although it allows for redesigning brand new processes, its focus appears to be most focused on working with an existing assembly line and finding ways to reduce its

***Table* 2.2** Process improvement methodologies and their areas of focus

| Methodology | Addresses | Solution set |
|---|---|---|
| BPR | Ineffective, inefficient processes | Create a better process, typically by radical redesign |
| Lean | Waste in the value stream | Identify wasted steps in the value stream, and where possible eliminate them |
| Six Sigma | Errors and variability of outputs | Identify causes of errors and variable outputs, often using statistical control techniques, and find ways to control for them |

inefficiencies. This approach would make sense for organizations which have established processes/value streams where there is a goal to make those processes/value streams more efficient.

Six Sigma was developed from a perspective of statistical control of industrial processes. At its heart, it focuses on variability in processes and error rates in production and seeks to control and limit variability and errors where possible. It asserts that variability and errors cost a company money, and learning to reduce these will increase profits. Similar to both BPR and Lean, it is dependent on top-level support to make the changes that will provide its benefits.

Whichever of these methods is selected to provide a more effective and efficient approach to doing business, it may be important to remember the lessons of the history of work since the beginning of the industrial revolution. We started with craftsmen satisfying the needs of a small base of customers. We then learned to increase productivity to satisfy the unmet demand of a much larger customer base, but in organizations that were focused inward on issues of productivity, not outward toward the customers. Now that we have reached a tipping point where supply can overtake demand, we need to again pay attention to customer needs for our organizations to survive and prosper. One of the process views of work may provide the means to do that.

## References

Drucker, P.F. (1993). *The Post Capitalist Society*. Harper Collins, New York.

Hammer, M. and Champy, J. (1993). *Reengineering the Corporation: A Manifesto for Business Revolution*. Harper Business, New York.

Heminger, A. (2014). Industrial revolution, customers, and process improvement. In A.B. Badiru (Ed.), *Handbook of Industrial and Systems Engineering*. CRC Press/Taylor & Francis Group, Boca Raton, FL.

Smith, A. (1776). *The Wealth of Nations*. Simon & Brown, London.

Womack, J.P. and Jones, D.T. (2003). *Lean Thinking: Banish Waste and Create Wealth in Your Corporation*. Free Press, New York.

# chapter three

# Human Factors in systems modeling

## Introduction

This chapter is based on Human Factors exposition from Resnick (2014). Human Factors is the mental or cognitive aspect of a system design while ergonomics is the physical aspect. Human Factors is a science that investigates human behavioral, cognitive, and physical abilities, and limitations in order to understand how individuals and teams will interact with products and systems (Resnick, 2014). Human Factors engineering is the discipline that takes this knowledge and uses it to specify, design, and test systems to optimize safety, productivity, effectiveness, and satisfaction. Human Factors is important to industrial and systems engineering because of the prevalence of humans within industrial systems. It is humans that, for the most part, are called on to design, manufacture, operate, monitor, maintain, and repair industrial systems. In each of these cases, Human Factors should be used to insure that the design will meet system requirements in performance, productivity, quality, reliability, and safety. This chapter presents an overview of Human Factors, how it should be integrated into the systems engineering process, and some examples from a variety of industries.

The importance of including Human Factors in systems design cannot be overemphasized. There are countless examples that illustrate the importance of Human Factors for system performance. Mackenzie (1994) found that in a survey of 1,100 computer-related fatalities between 1979 and 1992, 92% could be attributed to failures in the interaction between a human and a computer. The extent of the 1979 accident at the Three Mile Island nuclear power plant was largely due to Human Factors challenges (Bailey, 1996), almost resulting in a disastrous nuclear catastrophe. The infamous butterfly ballot problem in Florida in the 2000 U.S. presidential election is a clear example of an inadequate system interface yielding remarkably poor performance (Resnick, 2001). Websites such as www.baddesigns.com, thisisbroken.com, and others provide extensive listings of designs from everyday life that suffer from poor consideration of Human Factors. Neophytes often refer to Human Factors as common sense. However, the prevalence of poor design suggests that Human

Factors sense is not as common as one might think. The consequences of poor Human Factors design can be inadequate system performance, reduced product sales, significant product damage, and human injury.

This chapter provides an overview of Human Factors and is intended to support the effective design of systems in a variety of work domains, including manufacturing, process control, transportation, medical care, and others. Contents include some of the principal components of Human Factors analysis that must be addressed in any systems design and the benefits of integrated effective Human Factors. Also provided is the description of a conceptual model of human information processing, which addresses how each aspect affects performance. An example is provided for each one that illustrates the design challenges for Human Factors and how they can be overcome. The chapter also describes two important consequences of design: the ability of humans to learn from their experience, and the likelihood of error during system use.

## Elements of Human Factors

In order to facilitate the design of effective systems, Human Factors must adopt a holistic perspective on human–system interaction. Systems engineers need to understand how people think, how these thoughts lead them to act, the results of these actions, and the reliability of the results of these actions. Thus, the following four elements should be considered: cognition, behavior, performance, and reliability.

## Cognition

A considerable body of Human Factors research has been dedicated to human cognition. It is critical for systems engineers to understand and predict how users will perceive the information that they receive during system use, how this information will be processed, and the nature of users' resulting behavior and decisions. Situation awareness (SA) (Endsley, 2000a) refers to the extent to which a user has perceived and integrated the important information in the world and can project that information into the future to make predictions about system performance.

Consider the case of an air-traffic controller who needs to monitor and communicate simultaneously with several aircraft to ensure they all land safely. This job requires the controller to develop a composite mental model of the location and direction of each aircraft so that when new information appears, he or she can quickly decide on an appropriate response. The design of the system interface must anticipate this model so that information can be presented in a way that allows the controller to perceive it quickly and effectively integrate it into the mental model.

## *Behavior*

The actions taken by the human components of a system are often more difficult to predict than the mechanical or electrical components. Unlike machines, people behave based on experiences and beliefs that transcend the system, including factors such as corporate culture, personal goals, and past experience. It is critical for systems engineers to investigate the effects of these sources on behavior to ensure that the system will be successful.

For example, the Columbia Accident Investigation Board (CAIB) concluded that the accident causing the destruction of the space shuttle Columbia in 2003 was as much caused by the NASA organizational culture as it was by the foam that struck the orbiter. The CAIB report stated that systems were approved despite deviations in performance because of a past history of success (CAIB, 2003). At the consumer level, various factors may also be important. For instance, Internet retailers are interested in the factors that determine whether a consumer will purchase a product on the company's website. In addition to the design factors such as the site's menu design and information architecture, the user's past history at other websites can also affect his or her behavior on this site (Nielsen, 1999).

## *Performance*

Most systems depend not only on whether an action is completed, but also on the speed and accuracy with which the action is completed. Many factors affect user performance, such as the number of information sources that must be considered, the complexity of the response required, the user's motivation for performing well, and others (Sanders and McCormick, 1993).

Call-center operations are a clear example of the need to include Human Factors in design to achieve optimal performance (Kemp, 2001). Call-center software must complement the way that operators think about the task, or performance may be significantly delayed. The cost structure of call centers relies on most customer service calls being completed within seconds. Early versions of some customer relationship management (CRM) software required operators to drill down through ten screens to add a customer record. This design slowed the task considerably. However, labels that lead to strong path recognition can have as great an effect as path length on performance (Katz and Byrne, 2003). Trade-offs between path recognition strength and path length must be resolved in the information architecture of the system. It is critical that systems that rely on speed and accuracy of performance thoroughly integrate Human Factors into their designs.

## Reliability

Human Factors is also important in the prediction of system reliability. Human error is often cited as the cause of system failures (FAA, 1990; MacKenzie, 1994). However, the root cause is often traceable to an incompatibility between the system interface and human information processing. An understanding of human failure modes, the root causes of human error, and the performance and contextual factors that affect error probability and severity can lead to more reliable systems design.

Much of the research in human error has been in the domain of aerospace systems and control center operations (Swain and Guttmann, 1983). Also available are models that predict human errors in behavior, cognition, communication, perception, and others. Integrating human reliability models into the systems engineering process is essential.

## The benefits of Human Factors

There are many benefits that result from considering each of these four elements of Human Factors in systems design. The primary benefit is that the resulting system will be more effective. By accommodating the information-processing needs of the users, the system will better match the system requirements and will thus be more productive. Systems that incorporate Human Factors are also more reliable. Since human error is often the cause of system failure, reducing the likelihood of human error will increase the reliability of the system as a whole.

Consideration of Human Factors also leads to cost reductions in system design, development, and production. When Human Factors are considered early in the design process, flaws are avoided and early versions are closer to the final system design. Rework is avoided, and extraneous features can be eliminated before resources are expended on developing them.

Human Factors also leads to reduced testing and quality assurance requirements. Problems are caught earlier and it becomes easier to prioritize what system components to modify. Systems that exhibit good Human Factors design reduce sales time and costs because they are easier to demonstrate, train, and set up in the field.

Finally, consideration of Human Factors leads to reduced costs for service and support. When systems are easy to use, there are fewer service calls and less need for ongoing training. The occurrence of fewer errors leads to reduced maintenance costs, fewer safety violations, and less frequent need for mishap/injury investigation.

## A Human Factors conceptual model

Behavior and performance emerge from the way the humans process information. Human information processing is generally conceptualized

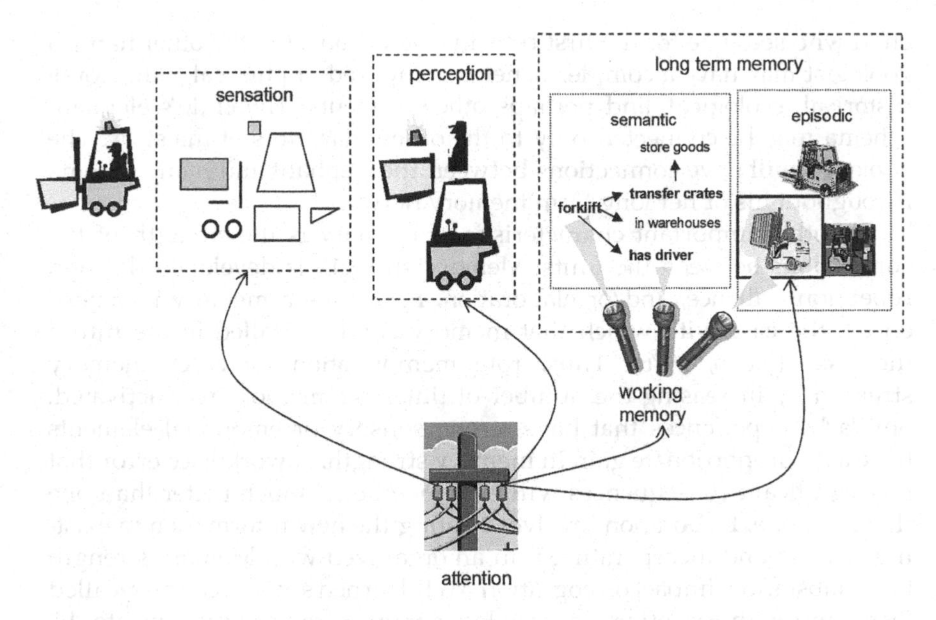


***Figure 3.1*** A conceptual model of human information processing.

using a series of processing stages (see Figure 3.1). It is important to keep in mind that these stages are not completely separate; they can work in parallel, and they are linked bi-directionally. A detailed discussion of the neurophysiology of the brain is beyond the scope of this chapter. But there is one underlying trait that is often overlooked. The human information processing system (the brain) is noisy, a fact that can lead to errors even when a particular fact or behavior is well known. On the positive side, this noise also enables greater creativity and problem-solving ability.

## *Long-term memory*

It refers to the composite of information that is stored in an individual's information processing system, the brain. It is composed of a vast network of interconnected nodes, a network that is largely hierarchical but that has many cross-unit connections as well. For example, a dog is an animal, but also lives (usually) in a house.

The basic unit of memory has been given many names, but for the purposes of this chapter will be called the *cell assembly* (Hebb, 1955). A cell assembly is a combination of basic attributes (lines, colors, sounds, etc.) that become associated because of a history of being activated simultaneously. Cell assemblies are combined into composites called schema. The size and complexity of the schema depend on the experience of the individual. The schema of an elephant will be very simple for a 4-year-old child who sees one for the first time in a storybook. On the other hand, a zoologist may have a complex schema composed of physical, behavioral, historical, ecological, and perhaps other elements. The child's elephant schema may be connected only to the other characters of the story. The zoologist will have connections between the elephant and many schema throughout his or her long-term memory network.

Another important characteristic of memory is the strength of the connections between the units. Memory strength is developed through repetition, salience, and/or elaboration. Each time a memory is experienced, the ease with which that memory can be recalled in the future increases (Hebb, 1976). Thus, rote memorization increases memory strength by increasing the number of times the memory was activated. Similarly, experiences that have strong sensory or emotional elements have a disproportionate gain in memory strength. A workplace error that has significant consequences will be remembered much better than one that has none. Elaboration involves relating the new information to existing schema and incorporating it in an organized way. Memory strength has a substantial impact on cognition. Well-learned schema can be recalled faster and with less effort because less energy is required to activate the stronger connections.

### *Types of long-term memory*

Long-term memory can be partitioned into categories such as episodic, semantic, and procedural components (Tulving, 1989). Episodic memory refers to traces that remain from the individual's personal experiences. Events from the past are stored as sub-sensory cell assemblies that are connected to maintain important features of the event but generalize less important features. Thus, an athlete's episodic memory of the championship game may include very specific and detailed visual traces of significant actions during the game. But less important actions may really be represented using statistical aggregates of similar actions that occurred over the athlete's total experience of games. These aggregates would be reconstructed during recall to provide the recall experience.

Semantic memory is composed of conceptual information, including knowledge of concept definitions, object relationships, physical laws, and similar non-sensory information (Lachman, Lachman, and Butterfield, 1979). Connections within semantic memory link concepts that are related or are used together. While the composition of semantic memory is not as structured or organized as an explicit semantic network (Figure 3.2), this is a reasonable simplification for the purposes of this chapter. There are also links between semantic memory and episodic memory. For example, the semantic memory of the meaning of "quality" may be linked to the episodic memory of a high-quality product.

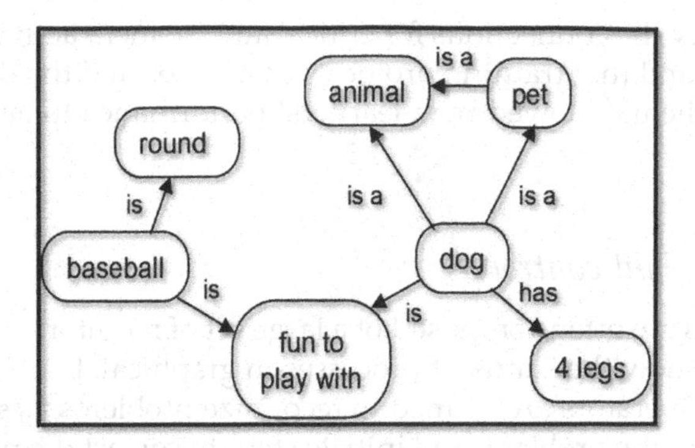


*Figure 3.2* A semantic network.

Procedural memory refers to the combination of muscle movements that compose a unitary action (Johnson, 2003). Procedural memories are often unconscious and stored as one complete unit. For example, a pianist may be able to play a complex piano concerto, but cannot verbally report what the fourth note is without first imagining the initial three notes. These automatic processes can take over in emergency situations when there is not enough time to think consciously about required actions.

Memories are not separated into distinct units that are clearly demarcated within the human information processing system. There are overlaps and interconnections within the memory structure that have both advantages and disadvantages for human performance. Creative problem-solving is enhanced when rarely used connections are activated to brainstorm for solution ideas. But this can lead to errors when random connections are assumed to represent fact or statistical associations are assumed to apply to inappropriate specific cases.

## *Implications for design*

The structure of long-term memory has significant implications for the design of industrial systems. Workers can master work activities faster when new processes match prior learning and overlap with existing schema. This will also reduce the probability that inappropriate schema will become activated in emergency situations, perhaps leading to errors. Similarly, terminology for labels, instructions, and information displays should be unambiguous to maximize speed of processing.

Simulator and field training can focus on developing episodic memories that bolster semantic memory for system architecture and underlying physical laws. Training breadth can expand the semantic schema to include a variety of possible scenarios while repetition can be used to solidify the connections for critical and frequent activities. Scielzo et al. (2002) found that training protocols that supported the development of accurate schema allowed new learners' performance to approach that of experts.

### *Case: power-plant control*

Control rooms are often composed of a large set of monitors, controls, and displays that show the status of processes in graphical, tabular, and digital readouts. Operators are trained to recognize problems as soon as they occur, diagnose the problem, and initiate steps to correct the problem. The design of this training is critical so that operators develop schema that effectively support problem-solving.

To maximize the ability of operators to identify major emergencies, training should include repeated simulations of these emergencies. According to Wickens et al. (2004), initial training leads to an accurate response for a given situation. But additional training is still necessary to increase the speed of the response and to reduce the amount of attention that is necessary for the emergency to be noticed and recognized. Thus, overlearning is important for emergency response tasks. With sufficient repetition, operators will have strong long-term memories for each emergency and can recognize them more quickly and accurately (Klein, 1993). They will know which combination of displays will be affected for each problem and what steps to take. Accurate feedback is critical to ensure that workers associate the correct responses with each emergency (Wickens et al., 2004). When errors are made, corrective feedback is particularly important so that the error does not become part of the learned response.

But problems do not always occur in the same way. In order for training to cover the variability in problem appearance, variation must be included in the training. Employees must be trained to recognize the diversity of situations that can occur. This will develop broader schema that reflect a deeper conceptual understanding of the problem states and lead to a better ability to implement solutions. Semantic knowledge is also important because procedures can be context-specific (Gordon, 1994). Semantic knowledge helps employees to adapt existing procedures to new situations.

Training fidelity is also an important consideration. The ecological validity of training environments has been shown to increase training transfer, but Swezey and Llaneras (1997) have shown that not all features of the real environment are necessary. Training design should include an evaluation of what aspects of the real environment contribute to the development of effective problem schema.

## Working memory

While the entire network of schema stored in long-term memory is extensive, it is impossible for an individual to recall simultaneously more than a limited set. Working memory refers to the set of schema that is activated at one point in time. A schema stored in long-term memory reverberates due to some input stimulus and can remain activated even when the stimulus is removed (Jones and Polk, 2002). Working memory can consist of a combination of semantic, episodic, and procedural memories. It can be a list of unrelated items (such as a shopping list), or it can be a situational model of a complex environment.

The size of working memory has been the focus of a great body of research. Miller's (1956) famous study that reported a span of 7 ± 2 is widely cited. However, this is an oversimplification. It depends on the size and complexity of the schema being activated. Many simple schemas (such as the single digits and letters used in much of the original psychology research) can be more easily maintained in working memory compared to complex schema (such as mental models of system architecture) because the amount of energy required to activate a complex schema is greater. Experience is also a factor. Someone who is an expert in aerospace systems has stronger aerospace systems schema, and therefore can recall schema related to aerospace systems with less effort because of this greater strength. More schemas can thus be maintained in working memory.

The size of working memory also depends on the ability of the worker to combine information into chunks (Wickens et al., 2004). Chunks are sets of working memory units that are combined into single units based on perceptual or semantic similarity. For example, mnemonics enhance memory by allowing workers to remember a single acronym, such as SEARCH for Simplify, Eliminate, Alter sequence, Requirements, Combine operations, and How often for process improvement brainstorming (from Konz and Johnson, 2000), more easily than a list of items.

How long information can be retained in working memory depends on the opportunity for workers to subvocally rehearse the information (Wickens et al., 2004). Without rehearsal, information in working memory is lost rapidly. Thus, when working memory must be heavily used, distractions must be minimized and ancillary tasks that also draw on this subvocalization resource must be avoided.

The similarity of competing information in working memory also affects the reliability of recall. Because working memory exists in the auditory modality, information that sounds alike is most likely to be confused (Wickens et al., 2004). The working memory requirements for any concurrent activities must be considered to minimize the risk of interference.

## *Implications for design*

A better understanding of working memory can support the development of more reliable industrial systems. The amount of information required to complete work activities should be considered in relation to the working memory capacities of the workers. Norman (1988) describes two categories of information storage. Information in the head refers to memory and information in the world refers to labels, instructions, displays, and other physical devices. When the amount of information to complete a task exceeds the capacity of working memory, it must be made available in the physical world, through computer displays, manuals, labels, and help systems. Of course, accessing information in the world takes longer than recalling it from memory, so this time must be considered when evaluating system performance requirements. This trade-off can also affect accuracy, as workers may be tempted to use unreliable information in working memory to avoid having to search through manuals or displays for the correct information (Gray and Fu, 2004).

Similarly, when information must be maintained in working memory for a long period of time, the intensity can fall below the threshold required for reliable recall. Here too, important information should be placed in the physical world. Interfaces that allow workers to store preliminary hypotheses and rough ideas can alleviate the working memory requirements and reduce the risk of memory-related errors. When information must be maintained in working memory for extended periods, the worker must be allowed to focus on rehearsal. Any other tasks that require the use of working memory must be avoided. Distractions that can interfere with working memory must be eliminated.

Training can also be used to enhance working memory. Training modules can be used to strengthen workers' conception of complex processes, and thus can reduce the working memory required to maintain it during work activities. This would allow additional information to be considered in parallel.

## *Case: cockpit checklists*

Degani and Wiener (1993) describe cockpit checklists as a way to provide redundancy in configuring an aircraft and reduce the risk of missing a step in the configuration process. Without checklists, aircraft crews would have to retrieve dozens of configuration steps from long-term memory and maintain in working memory whether each step had been completed for the current flight. Checklists reduce this memory load by transferring the information into the world. Especially in environments with frequent interruptions and distractions, the physical embodiment of a procedure can ensure that no steps are omitted because of lapses in working memory.

## *Sensation*

Sensation is the process through which information about the world is transferred to the brain's perceptual system through sensory organs such as the eyes and ears (Bailey, 1996). From a systems design perspective, there are three important parameters for each dimension that must be considered: sensory threshold, difference threshold, and stimulus–response ratio.

The sensory threshold is the level of stimulus intensity below which the signal cannot be sensed reliably. The threshold must be considered in relation to the work environment. In the visual modality, there are several important stimulus thresholds. For example, in systems that use lights as warnings, indicators and displays need to have a size and brightness that can be seen by workers at the appropriate distance. These thresholds were determined in ideal environments. When environments are degraded because of dust or smoke, or workers are concentrating on other tasks, the thresholds may be much higher. In environments with glare or airborne contaminants, the visual requirements will change.

Auditory signals must have a frequency and intensity that workers can hear, again at the appropriate distance. Workplaces that are loud or where workers will be wearing hearing protection must be considered. Olfactory, vestibular, gustatory, and kinesthetic senses have similar threshold requirements that must be considered in system design.

The difference threshold is the minimum change in stimulus intensity that can differentiated; this is also called the "just noticeable difference" (Snodgrass, Levy-Berger, and Haydon, 1985). This difference is expressed as a percent change. For example, a light must be at least 1% brighter than a comparison for a person to be able to tell that they are different. On the other hand, a sound must be 20% louder than a comparison for a person to perceive the difference.

The difference threshold is critical for the design of systems when multiple signals must be differentiated. When different alarms are used to signal different events, it is critical that workers be able to recognize the difference. When different-sized connectors are used for different parts of an assembly, workers need to be able to distinguish which connector is the correct one. Although there has been little research in this area, it is likely that there is a speed/accuracy trade-off with respect to difference thresholds. When workers are forced to act quickly, either because of productivity standards or in an emergency situation, even higher differences may be required for accurate selection.

The third dimension is the stimulus–response ratio. The relationship between the increase in intensity in a sensory stimulus and the corresponding increase in the sensation of that intensity is an exponential function (Stevens, 1975). For example, the exponent for perception of load heaviness is 1.45, so a load that is 1.61 times as heavy as another load will be perceived as twice as heavy. Similarly, the exponent for the brightness of a light is 0.33, so a light has to be eight times as bright to be perceived as twice as bright. Predicting these differences in perception is critical so that systems can be designed to minimize human error in identifying and responding to events.

### *Implications for design*

To maximize the reliability with which important information will reach workers, work environment design must consider sensation. The work environment must be designed to maximize the clarity with which workers can sense important sources of information. Lighting must be maintained to allow workers to see at requisite accuracy levels. Effective choice of color for signs and displays can maximize contrast with backgrounds and the accuracy of interpretation. Background noise can be controlled to allow workers to hear important signals and maintain verbal communication. The frequency and loudness of auditory signals and warnings can be selected to maximize comprehension. Location is also important. Key sources of visual information should be placed within the worker's natural line of sight.

### *Case: industrial dashboards*

Designing system interfaces to support complex decision-making, such as with supply chain, enterprise and executive dashboards, requires a focus on human sensory capabilities (Resnick, 2003). Display design requires selecting among digital, analog, historical, and other display types (Hansen, 1995; Overbye et al., 2002). The optimal design depends on how often the data change and how quickly they must be read.

The salience of each interface unit is also critical to insure that the relevant ones attract attention from among the many others on the display (Bennett and Flach, 1992). A variety of techniques can be used in industrial dashboards to create salience, such as brightness, size, auditory signals, or visual animation. The design should depend on the kinds of hardware on which the system will be implemented. For example, when systems will be accessed through handheld or notebook computers, the display size, and color capabilities will be limited and these limitations must be considered in the display design.

## *Perception*

As these basic sensory dimensions are resolved, the perceptual system tries to put them together into identifiable units. This is where each sensation is assigned to either an object or the background. If there are several objects, the sensations that compose each of them must be separated. There is a strong interaction here with long-term memory. Objects with strong long-term memory representations can be recognized faster and more reliably because less energy is required to activate the corresponding schema. On the other hand, objects that have similar features to different but well-known objects are easily misidentified as these objects. This is called a capture error because of the way the stronger schema "captures" the perception and becomes active first.

There is also an interaction with working memory. Objects that are expected to appear are also recognized faster and more reliably. Expectations can be described as the priming of the schema for the object that is expected. Energy is introduced into the schema before the object is perceived. Thus, less actual physical evidence is needed for this schema to reach its activation threshold. This can lead to errors when the experienced object is not the one that was expected but has some similarities.

## *Implications for design*

The implications of perception for industrial systems design are clear. When designing work objects, processes, and situations, there is a trade-off between the costs and benefits of similarity and overlap. When it is important that workers are able to distinguish objects immediately, particularly in emergency situations, overlap should be minimized. Design efforts should focus on the attributes that workers primarily use to distinguish similar objects. Workers can be trained to focus on features that are different. When object similarity cannot be eliminated, workers can be trained to recognize subtle differences that reliably denote the object's identity.

It is also important to control workers' expectations. Because expectations can influence object recognition, it is important that they reflect the true likelihood of object presence. This can be accomplished through situational training. If workers know what to expect, they can more quickly and accurately recognize objects when they appear. For those situations where there is too much variability for expectations to be reliable, work procedures can include explicit re-checking of the identity of objects where correct identification is critical.

## *Case: in-vehicle navigation systems*

In-vehicle navigation systems help drivers find their way by showing information on how to travel to a programmed destination. These systems can vary greatly in the types of information that they provide and the way in which the information is presented. For example, current systems can present turn-by-turn directions in the visual and/or auditory modalities, often adjusted according to real-time traffic information. These systems can also show maps that have the recommended route and traffic congestion highlighted in different colors.

There are many advantages provided by these systems. In a delivery application, optimization software can consider all of the driver's remaining deliveries and current traffic congestion to compute the optimal order to deliver the packages. For many multi-stop routes, this computation would exceed the driver's ability to process the information. Including real-time traffic information also enhances the capabilities of the system to select the optimal route to the next destination (Khattak, Kanafani, and Le Colletter, 1994).

A challenge for these systems is to provide this information in a format that can be quickly perceived by the driver. Otherwise, there is a risk that the time required for the driver to perceive the relevant information will require an extended gaze duration, increasing the likelihood of a traffic accident. Persaud and Resnick (2001) found that the display modality had a significant effect on decision-making time. Graphical displays, although the most common design, required the most time to parse. Recall scores were also lowest for graphical displays, possibly requiring the driver to look back at the display more often. This decrease in recognition speed can lead to the greater risk of a traffic accident.

## Attention and mental workload

Because only a small subset of long-term memory can be activated at any one time, it is important to consider how this subset is determined. Ideally, attention will be focused on the most important activities and the most relevant components of each activity, but the prevalence of errors that are the result of inappropriately focused attention clearly indicates that this is not always the case.

There are many channels of information, both internal and external, upon which attention can be focused. In most industrial settings, there can be visual and auditory displays that are designed specifically to present information to workers to direct their job activities. There are also informal channels in the various sights, sounds, smells, vibrations, and other sensory emanations around the workplace. Communication with other workers is also a common source of information. Additionally, there are internal sources of information in the memory of the individual. Episodic and semantic memories both can be a focus of attention. But it is impossible for workers to focus their attention on all of these channels at once.

It is also important to consider that attention can be drawn to channels that are relevant to the intended activities, but also to those that are irrelevant. Daydreaming is a common example of attention being focused on unessential information channels. Attention is driven in large part by the salience of each existing information channel. Salience can be defined as the attention-attracting properties of an object. It can be derived based on the intensity of a channel's output in various sensory modalities (Wickens and Hollands, 2000). For example, a loud alarm is more likely to draw attention than a quiet alarm. Salience can also be based on the semantic importance of the channel. An alarm that indicates a nuclear accident is more likely to draw attention than an alarm signaling lunchtime. Salience is the reason that workers tend to daydream when work intensity is low, such as long-duration monitoring of displays (control center operators, air travel, and security). When nothing is happening on the display, daydreams are more interesting and draw away the worker's attention. If something important happens, the worker may not notice.

If humans had unlimited attention, then we could focus on all possible information sources, internal and external. However, this is not the case; there is a limited amount of attention available. The number of channels on which attention can be focused depends on the complexity of each channel. One complex channel, such as a multifunction display, may require the same amount of attention as several simple channels, such as warning indicators.

Another important consideration is the total amount of attention that is focused on an activity and how this amount varies over time. This *mental workload* can be used to measure how busy a worker is at any given time, to determine if any additional tasks can be assigned without degrading performance, and to predict whether a worker could respond to unexpected events. Mental workload can be measured in several ways, including the use of subjective scales rated by the individual doing the activity or physiologically by measuring the individual's heart rate and/or brain function. A great deal of research has shown that mental workload must be maintained within the worker's capability, or job performance will suffer in domains such as air-traffic control (Lamoreux, 1997), driving a car (Hancock et al., 1990), and others.

## *Implications for design*

There are many ways to design the work environment to facilitate the ability of the worker to pay attention to the most appropriate information sources. Channels that are rarely diagnostic should be designed to have low salience. On the other hand, important channels can be designed to have high sensory salience through bright colors or loud auditory signals. Salient auditory alerts can be used to direct workers' attention toward key visual channels. Workers should be trained to recognize diagnostic channels so that they evoke high semantic salience.

Mental workload should also be considered in systems design. Activities should be investigated to ensure that peak levels of workload are within workers' capabilities. Average workload should not be so high as to create cumulative mental fatigue. It should also not be so low that workers are bored and may miss important signals when they do occur.

### *Case: warnings*

A warning is more than a sign conveying specific safety information. It is any communication that reduces risk by influencing behavior (Laughery and Hammond, 1999). One of the most overlooked aspects of warning design is the importance of attention. In a structured recall environment, a worker may be able to accurately recall the contents of a warning. However, if a warning is not encountered during the activity in which it is needed, it may not affect behavior because the worker may not think of it at the time when it is needed. When the worker is focusing on required work activities, the contents of the warning may not be sufficiently salient to direct safe behavior (Wogalter and Leonard, 1999). Attention can be attracted with salient designs, such as bright lights, sharp contrasts, auditory signals, large sizes, and other visualization enhancements.

Frantz and Rhoades (1993) reported that placing warnings in locations that physically interfered with the task could increase compliance even further. The key is to ensure that the warning is part of the attentional focus of the employee at the time it is needed and that it does not increase mental workload past the employees' capacity.

## *Situation awareness*

It is essentially a state in which an observer understands what is going on in his or her environment (Endsley, 2000a). There are three levels of SA: perception, comprehension, and projection. Perceptional SA requires that the observer know to which information sources attention should be focused and how to perceive these sources. In most complex environments, many information sources can draw attention. Dividing one's attention among all of them reduces the time that one can spend on critical cues and increases the chance that one may miss important events. This is Level 1 SA, which can be lacking when relevant information sources have low salience, are physically obstructed, are not available at needed times, when there are distractions, or when the observer lacks an adequate sampling strategy (Eurocontrol, 2003). The observer must be able to distinguish three types of information sources: those that must be examined and updated constantly, those that can be searched only when needed, and those that can be ignored.

Comprehension is the process of integrating the relevant information that is received into a cohesive understanding of the environment and retaining this understanding in memory for as long as it is needed. This includes both objective analysis and subjective interpretation (Flach, 1995). Comprehension can be compromised when the observer has an inadequate schema of the work environment or is over-reliant on default information or default responses (Eurocontrol, 2003).

Projection is when an observer can anticipate how the situation will evolve over time, can anticipate future events, and comprehends the implications of these changes. The ability to project supports timely and effective decision-making (Endsley, 2000a). One key aspect of projection is the ability to predict *when* and *where* events will occur. Projection errors can occur when current trends are under- or over-projected (Eurocontrol, 2003).

Endsley (2000a) cautions that SA does not develop only from official system interface sources. Workers can garner information from informal communication, world knowledge, and other unintended sources. SA is also limited by attention demands. When mental workload exceeds the observer's capacity, either because of an unexpected increase in the flow of information or because of incremental mental fatigue, SA will decline.

## *Implications for design*

Designing systems to maximize SA relies on a comprehensive task analysis. Designers should understand each goal of the work activity, the decisions that will be required, and the best diagnostic information sources for these decisions (Endsley, 2002).

It is critical to predict the data needs of the worker in order to insure that these are available when they are needed (Endsley, 2001). However, overload is also a risk because data must be absorbed and assimilated in the time available. To avoid overload, designers can focus on creating information sources that perform some of the analysis in advance and present integrated results to the worker. Displays can also be goal-oriented and information can be hidden at times when it is not needed.

It is also possible to design set sequences for the sampling of information sources into the work processes themselves. This can enhance SA because the mental workload due to task overhead is reduced. Workers should also be informed of the diagnosticity of each information source.

## *Case: air-traffic systems*

SA has been used in the investigation of air-traffic incidents and to identify design air-traffic control modifications that can reduce the likelihood of future incidents (Rodgers, Mogford, and Strauch, 2000). In the latter study, inadequate SA was linked to poor decision-making quality, leading both to minor incidents and major aircraft accidents. When air-traffic controllers are aware of developing error situations, the severity of the incident is reduced. The study identified several hypotheses to explain the loss of SA in both high-workload and low-workload situations. In high-workload conditions, operators had difficulty maintaining a mental picture of the air traffic. As the workload shifts down from high to low, sustained periods can lead to fatigue-induced loss of SA. The evaluation of air-traffic controller SA led to insights into the design of the radar display, communication systems, team coordination protocols, and data-entry requirements.

SA has also been used in the design stage to evaluate competing design alternatives. For example, Endsley (2000b) compared sensor hardware, avionics systems, free flight implementations, and levels of automation for pilots. These tests were sensitive, reliable, and were able to predict the design alternative that achieved the best performance.

## Decision-making

Decision-making is the process of selecting an option based on a set of information under conditions of uncertainty (Wickens et al., 2004). Contrary to the systematic way that deliberate decisions are made or programmed into computers, human decision-making is often unconscious, and the specific mechanisms are unavailable for contemplation or analysis by the person who made them. Environments with many interacting components, degrees of freedom, and/or unclear data sources challenge the decision-making process. Decision-making processes are affected by neurophysiological characteristics that are influenced by the structure of long-term memory and the psychological environment in which the decision is made. For experienced decision makers, decisions are situational discriminations (Dreyfus, 1997) where the answer is obvious without comparison of alternatives (Klein, 2000). There are two major types of decision-making situations: diagnosis and choice.

### Diagnosis

Diagnosis decisions involve evaluating a situation to understand its nature, and can be modeled as a pattern recognition process (Klein, 2000). Diagnosis describes decisions made in troubleshooting, medical diagnosis, accident investigation, safety behavior, and many other domains. The information that is available about the situation is compared to the existing schema in long-term memory, subject to the biasing effects of expectations in working memory. If there is a match, the corresponding schema becomes the diagnosis. For experts, this matching process can be modeled as a recognition-primed decision (Klein, 1993) whereby the environment is recognized as matching one particular pattern and the corresponding action is implemented.

The minimum degree to which the current situation must match an existing schema depends on the importance of the decision, the consequences of error, and the amount of time available. When the cost of searching for more information exceeds the expected benefits of that information, the search process stops (Marble, Medema, and Hill, 2002). For important decisions, this match threshold will be higher so that more evidence can be sought before a decision is made. This leads to more reliable and accurate decisions. However, it may still be the case that an observed pattern matches an existing schema immediately and a decision is made regardless of how important the decision is.

Under conditions of time pressure, there may not be sufficient time to sample enough information channels to reach the appropriate threshold. In these cases, the threshold must be lowered and decisions will be made based only on the information available (Ordonez and Benson, 1997). In these cases, individuals focus on the most salient source of information (Wickens and Hollands, 2000) and select the closest match based on whatever evidence has been collected at that point (Klein, 1993).

When the decision maker is an expert in the domain, this process is largely unconscious. The matched schema may be immediately apparent with no one-by-one evaluation of alternatives. Novices may have less well-structured schema and so the match will not be clear. More explicit evaluation may be required.

## *Choice*

In choice decisions, an individual chooses from a set of options that differ in the degree to which they satisfy competing goals. For example, when one is choosing a car, one model may have a better safety record and another may be less expensive. Neither is necessarily incorrect, although one may be more appropriate according to a specific set of optimization criteria.

When a person makes a decision, it is often based on an unconscious hybrid of several decision-making strategies (Campbell and Bolton, 2003). In the weighted average strategy, the score on each attribute is multiplied by the importance of the attribute, and the option with the highest total score is selected (Jedetski, Adelman, and Yeo, 2002). However, this strategy generally requires too much information processing for most situations and often doesn't match the desired solution (Campbell and Bolton, 2003). In the satisficing strategy, a minimum score is set for each attribute. The first option that meets all of these minima is selected (Simon, 1955). If none do, then the minimum of the least important attribute is relaxed and so on until an option is acceptable. In the lexicographic strategy, the option with the highest score on the single most important attribute is selected without regard for other attributes (Campbell and Bolton, 2003).

Table 3.1 illustrates a decision matrix that depicts the differences among these strategies. Using the weighted adding strategy, Option 1 would receive 173 points ($7 \times 8 + 3 \times 5 + 8 \times 9 + 5 \times 6$). Options 2 and 3 would receive 168 and 142, respectively. So Option 1 would be selected. On the other hand, the company may have satisficing constraints for attributes such as safety and value. A safety score less than 5 and a value score below 4 may be considered unacceptable regardless of the other attribute scores (eliminating Options 1 and 2 from consideration), resulting in the selection of Option 3. Finally, the company may choose to use a lexicographic strategy on value, selecting the option with the highest value regardless of all other attribute scores. In this case, Option 2 would be selected.

While the weighted adding strategy is often considered the most optimal, this is not necessarily the case. Some attributes, such as safety, should not be compensatory. Regardless of how fast, capable, reliable, or cost-effective a machine may be, risk to workers' safety should not be compromised. Lexicographic strategies may be justified when one attribute dominates the others, or the company does not have the time or resources to evaluate other attributes. For example, in an emergency situation, preventing the loss of life may dominate consideration of cost or equipment damage. The use of these strategies can be quite effective (Gigerenzer and Todd, 1999). And according to Schwartz (2004), benefits gained from making optimal decisions are often not worth the time and effort required.

Contrary to the systematic way the companies make official decisions, day-to-day decisions are often made with little conscious evaluation of the strategy (reference). As with diagnosis decisions, time pressure and decision importance influence the decision-making process. When faced with limited time, workers may be forced to use faster, simpler strategies such as the lexicographic strategy (Ordonez and Benson, 1997).

***Table 3.1*** Example decision-making matrix

| Attributes | Weights (out of 10) | Option 1 (scores) | Option 2 (scores) | Option 3 (scores) |
|---|---|---|---|---|
| Quality of output | 8 | 7 | 6 | 5 |
| Value | 5 | 3 | 9 | 5 |
| Safety | 9 | 8 | 4 | 7 |
| Durability | 6 | 5 | 7 | 4 |

## Decision-making heuristics

There are several decision-making heuristics that can reduce the information processing requirements and often reduce the time required to make a decision. However, these shortcuts can also bias the eventual outcome (Browne and Ramesh, 2002). These are often not consciously applied, so they can be difficult to overcome when they degrade decision-making accuracy and reliability.

- Anchoring: When an individual develops an initial hypothesis in either a diagnosis or choice decision, it is very difficult to switch to an alternative. Contrary evidence may be discounted.
- Confirmation: When an individual develops an initial hypothesis in either a diagnosis or choice decision, he or she will have a tendency to search for information that supports this hypothesis even when other channels may be more diagnostic.
- Availability: When searching for additional information, sources that are more easily accessed or brought to mind will be considered first, even when other sources are more diagnostic.
- Reliability: The reliability of information sources is hard to integrate into the decision-making process. Differences in reliability are often ignored or discounted.
- Memory limitations: Because of the higher mental workload required to keep many information sources in working memory simultaneously, the decision-making process will often be confined to a limited number of information sources, hypotheses, and attributes.
- Feedback: Similar to the confirmation bias, decision makers often focus on feedback that supports a past decision and discount feedback that contradicts past decisions.

## Implications for design

Human Factors can have a tremendous impact on the accuracy of decision-making. It is often assumed that normative decision-making strategies are optimal and that workers will use them when possible. However, neither of these is the case in many human decision-making situations. Limitations in information processing capability often force workers to use heuristics and focus on a reduced number of information sources. Competing and vague goals can reduce the applicability of normative decision criteria.

Workers can be trained to focus on the most diagnostic sources in each decision domain. If they are only going to use a limited number of sources, they should at least be using the most effective ones. Diagnostic sources also can be given prominent locations in displays or be the focus of established procedures.

The reliability of various information sources should be clearly visible either during the decision-making process or during training. Workers can be trained to recognize source reliability or to verify it in real time. Similarly, workers can be trained to recognize the best sources of feedback. In design, feedback can be given a more prominent position or provided more quickly.

To avoid anchoring and confirmation biases, Decision Support Systems (DSS) can be included that suggest (or require) workers to consider alternatives, seek information from all information sources, and include these sources in the decision-making process. At the least, a system for workers to externalize their hypotheses will increase the chance that they recognize these biases when they occur. However, the most successful expert systems are those that complement the human decision process rather than stand-alone advisors that replace humans (Roth, Bennett, and Woods, 1987).

In cases where decision criteria are established in advance, systems can be designed to support the most effective strategies. Where minimum levels of performance for particular criteria are important, the DSS can assist the worker in establishing the level and eliminating options that don't reach this threshold. The information-processing requirements of weighted adding strategies can be offloaded to DSS entirely to free the worker for information-collecting tasks for which he or she may be more suited.

## *Case: accident investigation*

Accident investigation and the associated root cause analysis can be fraught with decision-making challenges. Human error is often the proximate cause (Mullen, 2004) of accidents, but is much less often the root cause. During the accident investigation process, it is critical for investigators to explore the factors that led to the error and identify the design changes that will eliminate future risk (Doggett, 2004). However, this process engenders many opportunities for decision-making errors. Availability is usually the first obstacle. When an accident occurs, there is often high-visibility evidence that may or may not lead directly to the root cause. The CAIB report found that the root cause that ultimately led to the Columbia accident was not a technical error related to the foam shielding that was the early focus of the investigation, but rather was due to the organizational culture of NASA (CAIB, 2003). The confirmation bias can also challenge the investigation process. When investigators develop an initial hypothesis that a particular system component led to an accident, they may focus exclusively on evidence to confirm this component as the root cause rather than general criteria that could rule out other likely causes. This appeared in the investigations of the USS Vincennes incident in the Persian Gulf and the Three Mile Island nuclear power incident.

DSS can be used to remove a lot of the bias and assist in the pursuit of root causes. By creating a structure around the investigation, they can lead investigators to diagnostic criteria and ensure that factors such as base rates are considered. Roth et al. (2002) provide an overview of how DSS can reduce bias in decision-making. For example, DSS can inform users when the value for a particular piece of evidence falls outside a specified range. They can make confirming and disconfirming directions explicit and facilitate switching between them. But they warn that these systems can also introduce errors, such as by allowing drill-down into large data sources so that many data in one area are sampled without looking elsewhere. DSS can also exacerbate the availability bias by providing easy access to recent investigations.

## *Cognitive consequences of design*

### *Learning*

Every time an object is perceived, an event is experienced, a memory is recalled, or a decision is made, there are small, incremental changes in the structure of the human information processing system. Learning is very difficult when there is no prior experience to provide a framework (Hebb, 1955). This explains the power of analogies in early training. Later learning is a recombination of familiar patterns through the transition of general rules into automatic procedures and the generalization and specialization of these procedures as experience develops (Taatgen and Lee, 2003). The magnitude of the change depends on the salience of the experience and how well it matches existing schema.

When a human–system interaction is exactly the same as past experiences, there is very little learning because no new information is gained. The only result is a small strengthening of the existing schema. It is unlikely that the worker will develop a strong episodic memory of the event at all. When a human–system interaction is radically different from anything that has been experienced before, a strong episodic memory may be created because of the inherent salience of confusion and possible danger. But the event will not be integrated into the semantic network because it does not correspond to any of the existing schema – there is nowhere to "put" it. Maximum learning occurs when a human–system interaction mirrors past experience but has new attributes that make sense, i.e., the experience can be integrated into the conceptual understanding of the system.

### *Implications for design*

Training programs should always be designed based on an analysis of the workers' existing knowledge. Training of rote procedures where there will be no variability in workers' actions should be approached differently than training for situations where workers will be required to recognize and solve problems. In a study of novice pilot training, Fiore, Cuevas, and Oser (2003) found that diagrams and analogical representations of text content facilitate learning of procedures that require knowledge elaboration, but not on recognition or declarative rote memorization.

A better understanding of human learning mechanisms can also facilitate the development of experiential learning that workers gain on the job. System interfaces can be structured to maximize experiential learning by providing details that help employees develop accurate schema of the problem space. Over time, repeated exposure to this information can lead to more detailed and complex schema that can facilitate more elaborate problem solving. A cognitive analysis of the task requirements and possible situations can lead to a human–system interface that promotes long-term learning.

## Error

Human behavior is often divided into three categories: skill-based, rule-based, and knowledge-based (Rasmussen, 1993). In skill-based behavior, familiar situations automatically induce well-practiced responses with very little attention. In deterministic situations with a known set of effective responses, simple IF-THEN decision criteria lead to rule-based behaviors. Knowledge-based behaviors are required in unfamiliar or uncertain environments where problem solving and mental simulation is required.

Each of these behavior types is associated with different kinds of errors (Reason, 1990). With skill-based behavior, the most common errors are related to competing response schema. Skill-based behaviors result from a strong schema that is associated repeatedly with the same response. When a new situation shares key attributes with this strong schema, the old response may be activated in error. Because skill-based behaviors require little attention, the response is often completed before the error is noticed. In these cases, expertise can actually hurt performance accuracy. Unless there is salient feedback, the error may not be noticed and there will be no near-term recovery from the error.

Rule-based behaviors can lead to error when a rule is erroneously applied, either because the situation was incorrectly recognized or because the rule is inappropriately generalized to similar situations. Rule-based behavior is common with novices who are attempting to apply principles acquired in training. Because rule-based behavior involves conscious attention, the error is likely to be noticed, but the employee may not know of a correct response to implement.

Knowledge-based errors occur when the employee's knowledge is insufficient to solve a problem. Knowledge-based behavior is most likely to result in error because it is the type of behavior most often used in uncertain environments. When an employee is aware that his/her schema is not sophisticated enough to predict how a system will respond, he/she may anticipate a high likelihood of error and specifically look for one. This increases the chance that errors will be noticed and addressed.

### *Implications for design*

If system designers can anticipate the type(s) of behavior that are likely to be used with each employee–system interaction, steps can be taken to minimize the probability and severity of errors that can occur. For example, when skill-based behavior is anticipated, salient feedback must be designed into the display interface to ensure that employees will be aware when an error is made. To prevent skill-based errors from occurring, designers can make key attributes salient so that the inappropriate response will not be initiated.

To prevent rule-based errors, designers should ensure that the rules taught during training match the situations that employees will encounter when they are interacting with the system later. The triggers that indicate when to apply each rule should be made explicit in the system interface design. Signals that indicate when existing rules are not appropriate should be integrated into the interface design.

For complex systems or troubleshooting scenarios, when knowledge-based behavior is likely, errors can only be minimized when employees develop effective schema of system operators or when problem-solving activities are supported by comprehensive documentation and/or expert systems. Training should ensure that employees are aware of what they know and what they don't know. Employee actions should be easy to reverse when they are found to be incorrect.

## *Summary*

Humans interact with industrial systems throughout the system lifecycle. By integrating Human Factors into each stage, the effectiveness, quality, reliability, efficiency, and usability of the system can be enhanced. At the requirements stage, it is critical for management to appreciate the complexity of human–system interaction and allocate sufficient resources to ensure that Human Factors requirements are emphasized. During design, Human Factors should be considered with the earliest design concepts to maximize the match between human capabilities and system operations. As the system develops, Human Factors must be applied to control and display design and the development of instructions and training programs. Maintenance operations should also consider Human Factors to ensure that systems can be preserved and repaired effectively. Human Factors is also critical for human error analysis and accident investigation.

This chapter has presented a model of human information processing that addresses most of the relevant components of human cognition. Of course, one chapter is not sufficient to communicate all of the relevant Human Factors concepts that relate to the system lifecycle. But it does provide a starting point for including Human Factors in the process.

In addition to describing the critical components of human cognition, this chapter has described some of the implications of human cognition on system design. These guidelines can be applied throughout systems design. The specific cases are intended to illustrate this implementation in a variety of domains. As technology advances and the nature of human–system interaction changes, research will be needed to investigate specific human–system interaction effects. But an understanding of the fundamental nature of human cognition and its implications for system performance can be a useful tool for the design and operation of systems in any domain.

## References

Bailey, R. W. (1996). *Human Performance Engineering*, Third Edition. Prentice Hall, Upper Saddle River, NJ.

Bennett, K. B. and Flach, J. M. (1992). Graphical displays: Implications for divided attention, focused attention, and problem solving, *Human Factors* 34(5), 513–533.

Browne, G. J. and Ramesh, V. (2002). Improving information requirements determination: A cognitive perspective, *Information & Management* 39, 625–645.

CAIB. (2003). The Columbia accident investigation board final report. NASA. Available: www.caib.us (accessed August 4, 2004).

Campbell, G. E. and Bolton, A. E. (2003). Fitting human data with fast, frugal, and computable models of decision making. *Proceedings of the Human Factors and Ergonomics Society 47th Annual Meeting*, Human Factors and Ergonomics Society: Santa Monica, CA. pp. 325–329.

Degani, A. and Wiener, E. L. (1993). Cockpit checklists: Concepts, design, and use, *Human Factors* 35(2), 345–359.

Doggett, A. M. (2004). A statistical comparison of three root cause analysis tools, *Journal of Industrial Technology* 20(2), 2–9.

Dreyfus, H. L. (1997). Intuitive, deliberative, and calculative models of expert performance. In C. E. Zsambok and G. Klein (Eds.), *Naturalistic Decision Making*. Lawrence Erlbaum Associates, Mahwah, NJ.

Endsley, M. (2002). From cognitive task analysis to system design. Available: CTAResource.com.

Endsley, M. R. (2000a). Theoretical underpinnings of situation awareness: A critical review. In M. R. Endsley and D. J. Garland (Eds.), *Situation Awareness Analysis and Measurement*. Lawrence Erlbaum, Mahwah, NJ.

Endsley, M. R. (2000b). Direct measurement of situation awareness: Validity and use of SAGAT. In M. R. Endsley and D. J. Garland (Eds.), *Situation Awareness Analysis and Measurement*. Lawrence Erlbaum, Mahwah, NJ.

Endsley, M. R. (2001). Designing for situation awareness in complex systems. *Proceedings of the Second International Workshop of Symbiosis of Humans, Artifacts, and Environments*, Kyoto, Japan.

Eurocontrol. (2003). The development of situation awareness measures in ATM systems. European Organisation for the Safety of Air Navigation. Available: www.eurocontrol.int/humanfactors/docs/HF35-HRS-HSP-005-REP-01withsig.pdf (accessed June 18, 2004).

Federal Aviation Administration. (1990). Profile of operational errors in the national aerospace system. Technical Report. Washington, DC.

Fiore, S. M., Cuevas, H. M., and Oser, R. L. (2003). A picture is worth a thousand connections: The facilitative effects of diagrams on mental model development, *Computers in Human Behavior* 19, 185–199.

Flach, J. M. (1995). Situation awareness: Proceed with caution, *Human Factors* 37(1), 149–157.

Frantz, J. P. and Rhoades, T. P. (1993). A task-analytic approach to the temporal and spatial placement of product warnings, *Human Factors* 35, 719–730.

Gigerenzer, G. and Todd, P. (1999). *Simple Heuristics That Make Us Smart*. Oxford University Press, Oxford, UK.

Gordon, S. E. (1994). *Systematic Training Program Design*. Prentice Hall, Upper Saddle River, NJ.

Gray, W. D. and Fu, W. T. (2004). Soft constraints in interactive behavior: The case of ignoring perfect knowledge in-the-world for imperfect knowledge in-the-head, *Cognitive Science* 28, 359–382.

Hancock, P. A., Wulf, G., Thom, D., and Fassnacht, P. (1990). Driver workload during differing driving maneuvers, *Accident Analysis & Prevention* 22(3), 281–290.

Hansen, J. P. (1995). An experimental investigation of configural, digital, and temporal information on process displays, *Human Factors* 37(3), 539–552.

Hebb, D. O. (1955). *The Organization of Behavior*. John Wiley & Sons, New York.

Hebb, D. O. (1976). Physiological learning theory, *Journal of Abnormal Child Psychology* 4(4), 309–314.

Jedetski, J., Adelman, L., and Yeo, C. (2002). How web site decision technology affects consumers, *IEEE Internet Computing* 6(2), 72–79.

Johnson, A. (2003). Procedural memory and skill acquisition. In A. F. Healy, R. W. Proctor, and I. B. Weiner (Eds.), *Handbook of Psychology, Experimental Psychology*. John Wiley & Sons, Hoboken, NJ.

Jones, M. and Polk, T. A. (2002). An attractor network model of serial recall, *Cognitive Systems Research* 3, 45–55.

Katz, M. A. and Byrne, M. D. (2003). Effects of scent and breadth on use of site-specific search on e-commerce web sites, *ACM Transactions on Computer-Human Interaction* 10(3), 198–220.

Kemp, T. (2001). CRM stumbles amid usability shortcomings. *Internet Week Online*. Available: www.internetweek.com/newslead01/lead040601.htm (accessed August 4, 2004).

Khattak, A., Kanafani, A., and Le Colletter, E. (1994). Stated and reported route diversion behavior: Implications of benefits of advanced traveler information systems, *Transportation Research Record* 1464, 28.

Klein, G. (2000). *Sources of Power*. MIT Press, Boston, MA.

Klein, G. A. (1993). A recognition-primed decision (RPD) model of rapid decision making. In G. A. Klein, J. Orasanu, J. Calderwood, and D. MacGregor (Eds.), *Decision Making in Action: Models and Methods*. Ablex Publishing, Norwood, NJ.

Konz, S. and Johnson, S. (2000). *Work Design. Industrial Ergonomics*. Holcomb Hathaway, Scottsdale, AZ.

Lachman, R., Lachman, J. L., and Butterfield, E. C. (1979). Semantic memory. *Cognitive Psychology and Information Processing*. John Wiley & Sons, Hoboken, NJ.

Lamoreux, T. (1997). The influence of aircraft proximity data on the subjective mental workload of controllers on the air traffic control task, *Ergonomics* 42(11), 1482–14591.

Laughery, K. R. and Hammond, A. (1999). Overview. In M. S. Wogalter, D. M. DeJoy, and K. R. Laughery (Eds.), *Warnings and Risk Communication*. Taylor & Francis Group, Boca Raton, FL.

MacKenzie, D. (1994). Computer-related accidental death: An empirical exploration, *Science and Public Policy* 21(4), 233–248.

Marble, J. L., Medema, H. D., and Hill, S. G. (2002). Examining decision-making strategies based on information acquisition and information search time. *Proceedings of the Human Factors and Ergonomics Society 46th Annual Meeting*, Human Factors and Ergonomics Society: Santa Monica, CA.

Miller, G. A. (1956). The magical number seven, plus or minus two, *Psychological Review* 63, 81–97.

Mullen, J. (2004). Investigating factors that influence individual safety behavior at work, *Journal of Safety Research* 35, 275–285.

Nielsen, J. (1999). When bad designs become the standard. *Alertbox*. Available: www.useit.com/alertbox/991114.html (accessed August 4, 2004).

Norman, D. A. (1988). *The Design of Everyday Things*. Basic Books, New York.

Ordonez, L. and Benson, L. (1997). Decisions under time pressure: How time constraint affects risky decision making, *Organizational Behavior and Human Performance* 71(2), 121–140.

Overbye, T. J., Sun, Y., Wiegmann, D. A., and Rich, A. M. (2002). Human factors aspects of power systems visualizations: An empirical investigation, *Electric Power Components and Systems* 30, 877–888.

Persaud, C. H. and Resnick, M. L. (2001). The usability of intelligent vehicle information systems with small screen interfaces. *Proceedings of the Industrial Engineering and Management Systems Conference*, Institute of Industrial Engineers, Norcross, GA.

Rasmussen, J. (1993). Deciding and doing: Decision making in natural contexts. In G. A. Klein, J. Orasanu, J. Calderwood, and D. MacGregor (Eds.), *Decision Making in Action: Models and Methods*. Ablex Publishing, Norwood, NJ.

Reason, J. (1990). *Human Error*. Cambridge University Press, Cambridge, UK.

Resnick, M. (2014). Human factors. In A. B. Badiru (Ed.), *Handbook of Industrial and Systems Engineering*. CRC Press, Boca Raton, FL.

Resnick, M. L. (2001). Task based evaluation in error analysis and accident prevention. *Proceedings of the Human Factors and Ergonomics Society 45th Annual Conference*, Human Factors and Ergonomics Society, Washington, DC.

Resnick, M. L. (2003). Building the executive dashboard. *Proceedings of the Human Factors and Ergonomics Society 47th Annual Conference*, Human Factors and Ergonomics Society, Washington, DC.

Rodgers, M. D., Mogford, R. H., and Strauch, B. (2000). Post hoc assessment of situation awareness in air traffic control incidents and major aircraft accidents. In M. R. Endsley and D. J. Garland (Eds.), *Situation Awareness Analysis and Measurement*. Lawrence Erlbaum, Mahwah, NJ.

Roth, E. M., Bennett, K. B., and Woods, D. D. (1987). Human interaction with an intelligent machine, *International Journal of Man-Machine Studies* 27, 479–525.

Roth, E. M., Gualtieri, J. W., Elm, W. C., and Potter, S. S. (2002). Scenario development for decision support system evaluation. *Proceedings of the Human Factors and Ergonomics Society 46th Annual Meeting*, Human Factors and Ergonomics Society: Santa Monica, CA.

Sanders, M. S. and McCormick, E. J. (1993). *Human Factors in Engineering and Design*, Seventh Edition. McGraw-Hill, New York.

Schwartz, B. (2004). *The Paradox of Choice*. Harper Collins, New York.

Scielzo, S., Fiore, S. M., Cuevas, H. M., and Salas, E. (2002). The utility of mental model assessment in diagnosing cognitive and metacognitive processes for complex training. *Proceedings of the Human Factors and Ergonomics Society 46th Annual Meeting*, Human Factors and Ergonomics Society: Santa Monica, CA.

Simon, H. A. (1955). A behavioral model of rational choice, *Quarterly Journal of Economics* 69, 99–118.

Snodgrass, J. G., Levy-Berger, G., and Haydon, M. (1985). *Human Experimental Psychology*. Oxford University Press, Oxford, UK.

Stevens, S. S. (1975). *Psychophysics*. John Wiley & Sons, Hoboken, NJ.

Swain, A. D. and Guttmann, H. E. (1983). *A Handbook of Human Reliability Analysis with Emphasis on Nuclear Power Plant Applications*. NUREG/CR-1278. USNRC, Washington, DC.

Swezey, R. W. and Llaneras, R. E. (1997). Models in training and instruction. In G. Salvendy (Ed.), *Handbook of Human Factors and Ergonomics*, Second Edition. John Wiley & Sons, Hoboken, NJ.

Taatgen, N. A. and Lee, F. J. (2003). Production compilation: A simple mechanism to model complex skill acquisition, *Human Factors* 45(1), 61–76.

Tulving, E. (1989). Remembering and knowing the past, *American Scientist* 77, 361–367.

Wickens, C. D. and Hollands, J. G. (2000). *Engineering Psychology and Human Performance*. Prentice Hall, Upper Saddle River, NJ.

Wickens, C. D., Lee, J. D., Liu, Y., and Gordon-Becker, S. E. (2004). *An Introduction to Human Factors Engineering*, Second Edition. Prentice Hall, Upper Saddle River, NJ.

Wogalter, M. S. and Leonard, S. D. (1999). Attention capture and maintenance. In M. S. Wogalter, D. M. DeJoy, and K. R. Laughery (Eds.), *Warnings and Risk Communication*. Taylor & Francis Group, Boca Raton, FL.

# chapter four

# Ergonomics in systems design and modeling

## Introduction

This chapter is based on Peacock (2014). Human Factors is the mental or cognitive aspect of a system design while ergonomics is the physical aspect. Design is the process of converting the voice of the customer into some product or process. The problem however is which customer's voice do designers listen to. Take the examples of a smart phone or a car or the process of getting money from the bank or the university you choose to attend. All of these have many customers (or stakeholders), each with different requirements. The end users are interested in effectiveness and ease of use of the product or process for their particular needs, which may differ. They will also be interested in cost. Others may emphasize safety and security. But there are other "customers" perhaps with different requirements. What about the line workers who assemble the product or the employees of the bank or university? There are also the managers and shareholders of the companies who manufacture the products or manage the processes. They are interested in sales and profits. The result is that design is a process of resolving many, sometimes conflicting requirements.

One way of addressing this problem of design is to look at the ergonomics of the design process. There are many stages in this process including concept development and selection through manufacturing process design and production, to sales, operations, and maintenance. Recently, design for disposal has become a concern for many products. Each stage has different customers. Surprisingly, much of this design process takes place around a table where the different stakeholders communicate the importance of their particular requirements. And this is where the fun starts and the ergonomics of process design can have its opportunity to shine.

The first step is to agree on the scope of the many requirements. In general, the purposes of all products, processes, and organizations are Effectiveness, Efficiency, Ease of Use, Elegance, Safety, Security, Sustainability, and Satisfaction (E4S4). These purposes are deliberately general and comprehensive, but they will include more specific and quantitative concepts in particular instances. For example, efficiency includes the optimal use of resources such as money, time, people, and materials and security includes process failure due to malicious or accidental acts of third parties. Most organizations will identify growth and stability as primary purposes, but in reality these are dependent on E4S4.

The next step is to agree on the language of communication. Perhaps the best language is an adaptation of Quality Function Deployment (QFD), which has been used in the automobile industry for a couple of decades. It is important to separate the concepts of requirements and specifications. Requirements are what the customers want, and specifications are what the designers and engineers need. Requirements reflect the use of the product or process by a customer in a context. Specifications are those numbers that accompany the lines on the engineering drawing or project plan. Car owners may want vehicles that are comfortable to drive on long journeys or they may require cell phones that can also be used for e-mail and to surf the web. The manufacturing employees want products that are easy to assemble and maintainers like easy access to their tasks.

## *The design process*

Design is the process of converting the voice of the customer into some product or process. The problem, however, is which customers' voice do designers listen to? Take the examples of a smart phone or a car or the process of getting money from the bank or the university you choose to attend. All of these products and services have many customers (or stakeholders), each with different requirements. The end users, who often have a choice among competitive products or services, are interested in effectiveness and ease of use for their particular needs. Some may also be interested in esthetics and cost while others may emphasize safety. Figure 4.1 shows examples used for illustration of the design and evaluation of products and services.

There are other "customers" such as the production line workers and service organization employees who have less choice; they will have

| **Processes with Requirements** | **Systems or Products with Specifications** |
| --- | --- |
| • Transportation | • Car, Train |
| • Communication | • Mobile Phone |
| • Education | • Course |
| • Remuneration | • ATM |
| • Recreation | • Game |
| • Control | • TV Remote Control |

***Figure 4.1*** Examples used for illustration of the design and evaluation of products and services.

different requirements, such as ease of access for assembly or to their computer workstation. There are also the managers and shareholders of the companies that manufacture the products or oversee the services; they are interested in quality, productivity, sales, and profits. Design, therefore, is a process of resolving many, sometimes conflicting, requirements.

## *Design purposes and outcomes*

A general model of product or service design objectives can be summarized as "E4S4" (Figure 4.2). Effectiveness, or quality, is a measure of how well the product meets customer requirements. Efficiency reflects the use of resources such as time, money, materials, energy, and people. Ease of use indicates the comfort and convenience of various users as they interact with the product or service. Elegance is the esthetic appeal of the product that reflects the emotional attachment with the user. Safety is a description of the hazards associated with product or service interaction, including barriers to failure and mitigation following product or service failure. Security of a product or service is the degree to which malicious or accidental misuse can be prevented. Satisfaction is a composite measure of the various customers' experience with the product or service. Finally, sustainability indicates the reliability of the outcomes of a product or service over time (reliability) and the degree to which the product or service can withstand unexpected and sometimes extreme contexts (resilience).

The effectiveness of a transportation system is a measure of whether or not the chosen product or service achieves the objectives of a particular "transaction" or journey. For example, the choice among walking, bicycling, driving a car, or flying in an airplane will depend on the distance of the journey – the transaction requirements. Given the selection of an effective system, such as a car, the customer may be concerned with the efficiency of this choice such as vehicle cost or fuel consumption. The ease of use criterion may relate to the vehicle information systems or the ease

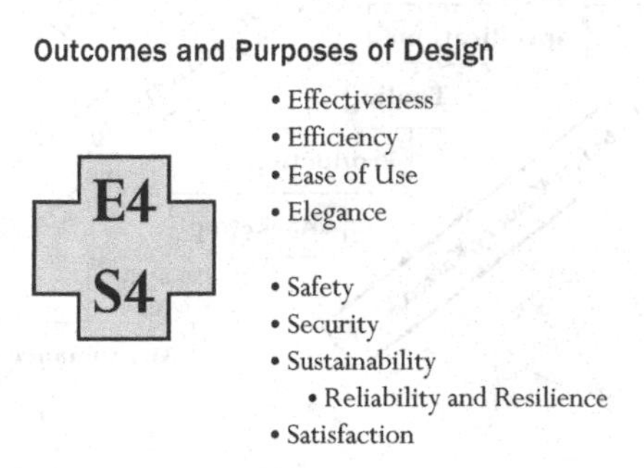

Outcomes and Purposes of Design
E4
S4
• Effectiveness
• Efficiency
• Ease of Use
• Elegance
• Safety
• Security
• Sustainability
• Reliability and Resilience
• Satisfaction


***Figure 4.2*** A comprehensive model for product or service design objectives.

of parking. The esthetic measure – "elegance" will indicate those subjective impressions of the vehicle's styling or nameplate. The safety objective may reflect the sophistication of the vehicle control features such as anti-lock brakes or reversing indicators; also, the mitigation features such as air bags and "friendly" dashboards may be included in the safety evaluation. Security in the case of a car may include locks and GPS tracking systems. Sustainability is measured by the evaluation of the vehicles' reliability and appeal over time. Resilience reflects the performance of the vehicle in extreme conditions including natural contexts such as off road routes and floods and human made challenges such as crashes and fuel shortages. Satisfaction is a composite measure of the total experience of the car user as reflected in owner surveys – customer loyalty to a brand will depend on the weighting of the various evaluation criteria, such as cost, styling, or safety features.

## *Life cycle*

One way of addressing the multiple purposes of design is to look at the ergonomics of the design process. There are many stages in this process (Figure 4.3) including concept development and selection through manufacturing process design and production, to sales, operations, and maintenance. Recently, design for disposal has become a concern for many products. Each stage has different customers and stakeholders. Many of the design decisions are made around a table where the different stakeholders communicate the importance of their particular requirements.

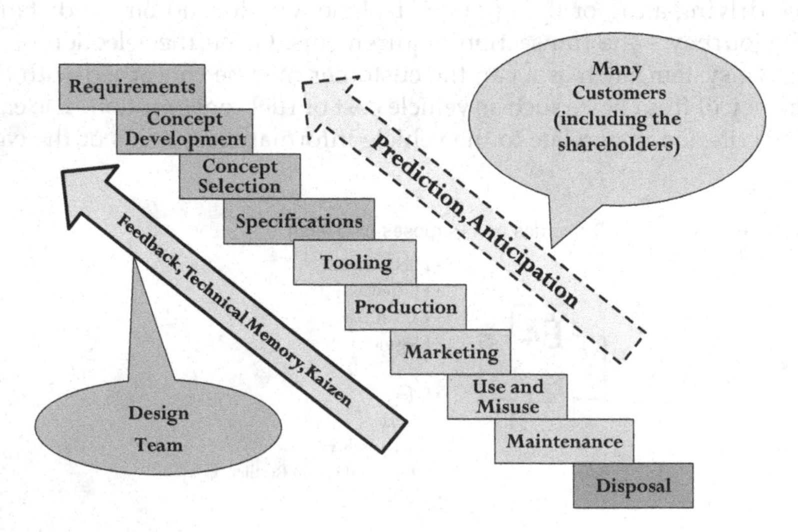


***Figure 4.3*** Product or service life cycle.

And this is where the fun starts and the ergonomics of process design can have its opportunity to shine.

The "cradle to grave" description of a product or service encompasses design, production, use, and eventual disposal. In a large organization, the marketing group will interpret the requirements articulated by the end user, sometimes with the help of an ergonomist. The Human Factors specialists will also work with other customers and stakeholders, such as manufacturing and maintenance, to extend these requirements. The final set of requirements involving various E4S4 criteria will inevitably be a compromise worked out among marketing, design, engineering, manufacturing, Human Factors, and cost departments with management oversight. Given these requirements, the design team will explore many concepts that conform to the different requirements to various degrees. Convergence on a single concept for development will inevitably be a compromise among the many customer and stakeholder biases.

For example, a major design decision for a mobile phone will be between hard and soft keys which imply both different technologies and different use of the product's interface "real estate" surface. The Human Factors specialist will evaluate effectiveness, efficiency, and customer satisfaction as reflected, for example, by the feedback provided to the user in time or light constrained operations. Given the very compact nature of this product, the ergonomist will assess the design of the production line layout, tools and the various jigs and fixtures needed to provide access and stability for assembly and inspection.

## *Structures, processes, and outcomes*

Donabedian (1998) described the health-care processes in terms of structures, processes, and outcomes (Figure 4.4). These concepts may also be applied to any complex entity such as a university where the structures include buildings, facilities, laboratories, and faculty members. Structures or entities are designed by reference to specifications. The processes include lectures and examinations and various management processes needed to assure that the institution runs smoothly. Generally, processes have requirements. The outcomes include employment of graduates and publications by faculty members. Generally, the structures are entities described by nouns and designed with reference to adjectives. Processes, on the other hand, are activities or verbs that are measured by reference to adverbs. The outcomes will generally be a change in the status (adjectives) of one or more of the input structures.

The contribution of the ergonomist is to help to translate requirements into specifications. In the case of a car, the requirement "comfortable to drive" is translated into the interior package dimensions, seat adjustment range, contours, and foam density. Other specifications may relate to the

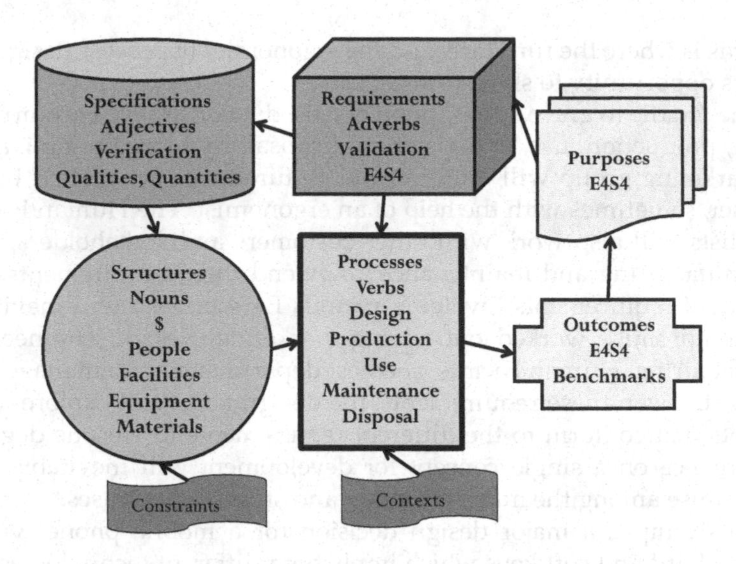


*Figure 4.4* Structures, processes, and outcomes.

driving environment and perhaps to the driver. Surfing the web can be achieved by specifying a command structure for hard keys or a menu process using icons and large buttons on a touch screen. The manufacturing design requirements of easy access may be achieved by a layering of modules and orientation of fasteners toward the assembler or maintainer. The process of QFD can be used to further analyze the relationships between requirements and specifications (Akao, 1990).

The evaluation or validation of processes will take place in a real world context or realistic simulations with representative users and transactions. Validation will address process purposes, requirements, and outcomes. For example, an ATM transaction may require multiple layers of security including temporary passwords that may confuse elderly users (Figure 4.5).

The customer requirements for getting money out of the bank include ease of use of the teller machine and security of the process. Both requirements may be achieved by specifying primary and secondary PIN designs that are easy to remember and hard to steal. This challenge indicates a trade-off between ease of use and security. The efficiency/effectiveness trade-off of a transaction will be reflected in the length of the encrypted password and the availability of "quick cash" options in common amounts. The ATM example is a simplified version of a web page where the number and complexity of choices and navigation through sequences of choices may be accomplished by such interactions as point and click or involve keyboard choices and inputs. User performance (effectiveness and efficiency) and satisfaction with web page transactions will be reflected by

*Figure 4.5* The ATM.

objective measures of speed and accuracy and subjective measures that are related to the transaction content.

Product design specifications are driven by process requirements, which in turn depend on intended outcomes. Whereas a requirement may be relative – the speed and accuracy of a text message on one mobile phone should be "better than" those of another – the design specifications must be objective and quantitative. A specification is absolute, perhaps with some tolerance: the font size should be at least 18 pts. There may also be trade-offs among multiple requirements. For example, the size of a key pad and hence the spacing of the keys will give rise to speed – accuracy trade-offs. Furthermore, these performance trade-offs may be preempted by a requirement to minimize the size of the overall device. Similar trade-offs between effectiveness and efficiency will arise with screen and associated symbol design. The requirements for the Holy Grail of the ideal communication devices (e.g., iPhone or iPad) will depend on particular usage and context of use criteria. Consequently, the specifications, such as screen/symbol and key size and spacing will derive from a compromise among performance requirements. It is usually an easy matter to verify that the design specifications have been met. However, where design constraints are encountered such as cost, weight or technology there may have to be trade-offs among the specifications with resulting implications for the outcomes.

In all these cases – cars, smart phones, bank account access, and website design – the role of the ergonomist is to work with the customers and the designers on the translation of requirements into specifications. The first task is to ensure that customers talk in terms of functions or activities and not the means of implementing these functions. Designers and engineers on the other hand should not try to dictate customer requirements, but rather understand what the customer wants, and the context of use of the product or service. The ergonomist, using a box of investigation, analysis, simulation, and experimentation tools, can contribute effectively to the task of translating the voice of the customer into the specifications of the product or service.

## Prediction

The main challenge for design is prediction (Figure 4.6). Seen as a control problem, design is about the development of structures to contribute to processes with desirable outcomes. Most "designs" have precedents and therefore can benefit from feedback; over time, designs are adapted to the context and the designers learn from their successes and failures. But progress involves uncertainty and risk, and technical memory is not always as effective as it should be. Also, the contexts and uncontrollable external forces are not always precisely predictable. Therefore, the designers must design for intended users, use and contexts, and foreseeable misuse and unexpected contexts. The product liability courts are full of cases where system failures resulted in unwanted outcomes. Also, design

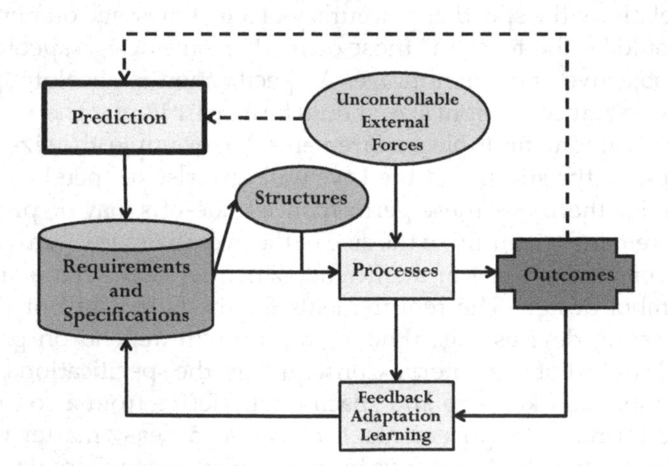


***Figure 4.6*** Anticipation, prediction, and control.

compromises may result in vulnerabilities to misuse or hostile contexts. A second challenge in prediction is that of user variability. Whereas the majority of experienced and trained users may use the product or process effectively, efficiently, and safely, novice and otherwise degraded users may misuse the product, perhaps with catastrophic results.

The transportation example will be used to illustrate these challenges of prediction in design. The designer of a car only has imperfect knowledge of the potential user. Furthermore, the car buyer may choose which car he or she buys and what journeys to make; they may also have exaggerated opinions regarding their driving abilities. Finally, the unpredictable context of other drivers, traffic, traffic regulations, roads, and weather add to the complexity of prediction, requirements, specifications, and outcomes. For example, one may attempt to develop the design specifications for a car aimed at the older driver. The requirements are derived from predictions regarding intended and unintended contexts and outcomes. Figure 4.7 shows some general examples of design types and a caricature of an "ideal" vehicle. The book shown in the figure provides extensive background on requirements and specifications for vehicle design (Peacock and Karwowski, 1993). The minivan has two advantages – ease of access and plenty of room for the grandchildren. Where mass and safety are concerned, the full-size pickup will provide good visibility and good

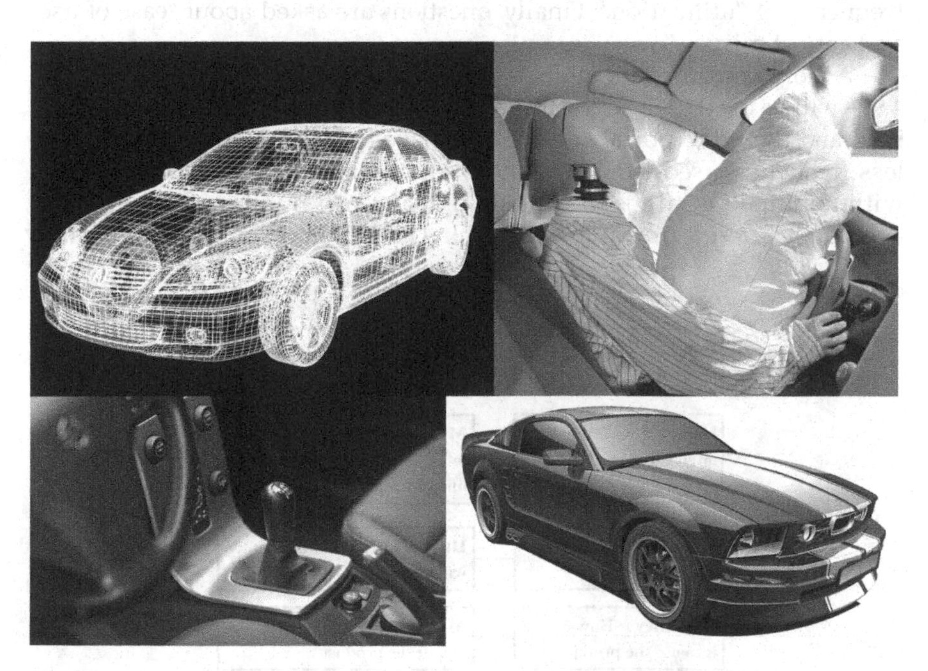

***Figure 4.7*** Automobile design. (© Can Stock Photo, Inc./rook, uatpl, lightpoet, and busja.)

performance in the event of a crash. If the elderly driver has accumulated sufficient wealth and wishes to relive his youth, then the Corvette may be the vehicle of choice, although entry, egress, and visibility may be a concern for the older driver, not to mention the response of the accelerator. The family sedan is perhaps the best compromise, especially if the physical, cognitive, and operational interfaces are geared toward the diminishing sensory and cognitive capacities of the older driver. If the driving environment is away from the busy freeways, the journeys are short and the weather is fine then a golf cart may suffice.

## *The 6Us and 2Ms*

An alternative way of analyzing product use with the purpose of predicting outcomes and therefore requirements is to use the 6Us and 2Ms template (Peacock and Resnick, 2010), for either descriptive or quantitative purposes (Figure 4.8). The first question is "how will the product or service be used?" Further depth to this question asks, "how useful is the product or service?" This second "utility" question may reflect criteria of efficiency and safety over and above the primary question of effectiveness. The next questions are about the intended user and possible misuser. Next come questions about the method and conditions of "usage" and the frequency of "utilization." Finally, questions are asked about "ease of use and ease of misuse."

A close cousin of the mobile phone – the TV remote control (Figure 4.9) – will be used to describe this analysis process. Both devices are handheld and therefore portable, they both have a plethora of frequently and less frequently used functions and buttons, and they both may be used with minimal visual feedback, although a wrong number or choice may eventually provide unwanted feedback. The TV remote control is typically

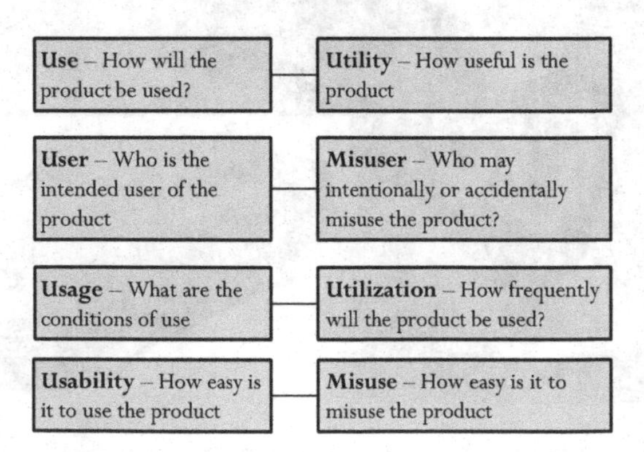


*Figure 4.8* The 6Us and 2Ms.

***Figure 4.9*** The TV remote control. (© Can Stock Photo, Inc./tshooter and kmitu.)

used to choose among many broadcast information and entertainment products. Its utility is really in the realm of efficiency – one does not have to move from the chair to search and select the desired program. Because of this effectiveness and efficiency, the device is used by the widest possible range of users – from the very young to the very old. The utilization or frequency of use can range from a few times a day to dozens of times because of this effectiveness and convenience. The usage or conditions of use are where the trouble starts. Typically, the device will be used with minimal environmental lighting. Misuse opportunities abound. A common misuse is to select the wrong device from the four or five that typically adorn coffee tables. The user and usage conditions, coupled with the large number of closely spaced and undifferentiated keys (from the tactile perspective) lead to a high probability of error. This basic shortcoming is compounded by the large number of features and choices, many of which are not used or rarely used by most users. A second design shortcoming is the lack of a simple accessible feature to help the user recover from an inadvertent error.

## *A case study in design – the Hong Kong Mass Transit Railway*

The Hong Kong Mass Transit Railway carries millions of passengers a day (Figure 4.10). For the passenger compartment, the principal passenger criteria are stability, safety, and mobility. Given the relatively short journeys, comfort is a lesser consideration. Journey time and cost are also key issues. The organization is interested in payload – passengers per square meter, and dwell time at stations – 25 passengers off and 25 passengers on the train in 25 seconds. Other considerations included emergency evacuation, maintenance, and resilience. The primary ergonomics opportunities

***Figure 4.10*** The Hong Kong Mass Transit Railway.

to fulfill these requirements were in the design of the inward facing bench seats and the arrangement of the horizontal rails and vertical poles.

The design process involved extensive use of physical mockups and usability trials to supplement the basic principles of anthropometry and biomechanics. Compromises had to be made between reach and fit – could the shorter passengers reach the high-level horizontal bar and would the taller passengers have sufficient head clearance? Given that standing passengers were intended to cluster around the vertical poles, these were located to allow the best possible access and mobility for passengers leaving or entering the train.

The design of train operator compartment ("operator" was preferred to "driver," given the substantial amount of automation) also had important ergonomics criteria such as control effectiveness, comfort, and stability; forward vision was also important. The travel times between stations were generally of the order of a few minutes; at each station, the operator was required to alight and check the status of passenger movement between the platform and the train, with the help of video cameras. Because of the constraints on the size of the operator's compartment, a full static seat was impractical. Consequently, a novel seat was designed which allowed the driver to choose between sitting, leaning on the folded seat or standing, depending on the segment duration. The control panel included a prominent emergency stop button and a handle that fulfilled acceleration (forward) and braking (backward) functions.

## *Qualitative and quantitative design tools*

The initial use of an ergonomics design tool is to serve as a checklist to ensure that the full spectrum of relevant questions is asked. Next, the tool must ensure sufficient depth and detail to facilitate the development of relevant requirements and clear specifications. Two such tools are described here – to facilitate the choices among design options at both the general and the detailed level. The first tool (Figure 4.11) lists the general E4S4 purposes with opportunities for more detailed criteria. It should be noted that the criteria may be differentially weighted and evaluated in a way that minor issues (tipping or halo factors) do not dominate the overall decisions.

The use of this worksheet may be either qualitative or quantitative. For convenience, simple cryptic words may be entered to summarize the performance of the alternative products or services on each of the criteria. There is a plethora of scoring rubrics – varying from simple pass/fail, through rankings and ratings to detailed measurements of times and costs, and composite customer scores.

| **Go to work – 20 miles, variable traffic and weather** | **Criterion** | **Walk** | **Bicycle** | **Motor bike** | **Bus** | **Taxi** | **Train** | **Car** |
|---|---|---|---|---|---|---|---|---|
| Effectiveness | | | | | | | | |
| Efficiency | Time, energy | | | | | | | |
| Ease of Use | Convenience | | | | | | | |
| Elegance | Status | | | | | | | |
| Safety | Cras protection | | | | | | | |
| Security | Locking | | | | | | | |
| Satisfaction | Composite | | | | | | | |
| Sustainability | Resilience | | | | | | | |
| **Score** | | | | | | | | |

***Figure 4.11*** An E4S4 evaluation sheet for a long journey to work.

| TV Remote Control | On TV | Single Device Minimal Functions | Single Device Multiple Functions | Multiple Devices Simple Functions | Multiple Devices Complex Functions |
|---|---|---|---|---|---|
| Use | | | | | |
| Utility | | | | | |
| User | | | | | |
| Usage | | | | | |
| Utilization | | | | | |
| Safety | | | | | |
| Security | | | | | |
| Satisfaction | | | | | |
| Sustainability | | | | | |
| **Score** | | | | | |

*Figure 4.12* 6Us and 2Ms evaluation of a TV remote control.

A second worksheet addresses the 6Us and 2Ms, again with the purpose of comparing a set of alternative designs (Figure 4.12).

This investigation method may be supplemented by the 5Ws and a How (Who, What, Where, When, Why, How) and the "5 Whys" in which each use evaluation is queried by asking "Why?" a series of times until a root cause of a problem or requirement is defined.

## *Concurrent engineering*

Concurrent engineering is the process of strategic overlap of design phases (product, manufacturing, production, etc.) to ensure that all customer requirements are considered as early as possible in the design process (Figure 4.13). Failure to do this can result in costly and time-consuming changes. Successful concurrent engineering results in a reduction of the total time between concept development and production. Also, the process facilitates effective trade-offs among the requirements of different customers.

Car design from scratch can take up to 5 years, by which time the faster competitors have their vehicles on the road and the design may contain aging technology. The concurrent engineering process can greatly reduce this "time to market" challenge. A complementary strategy is to use proven technology, such as engine and chassis modules and fasteners, and differentiate the product by cosmetic changes.

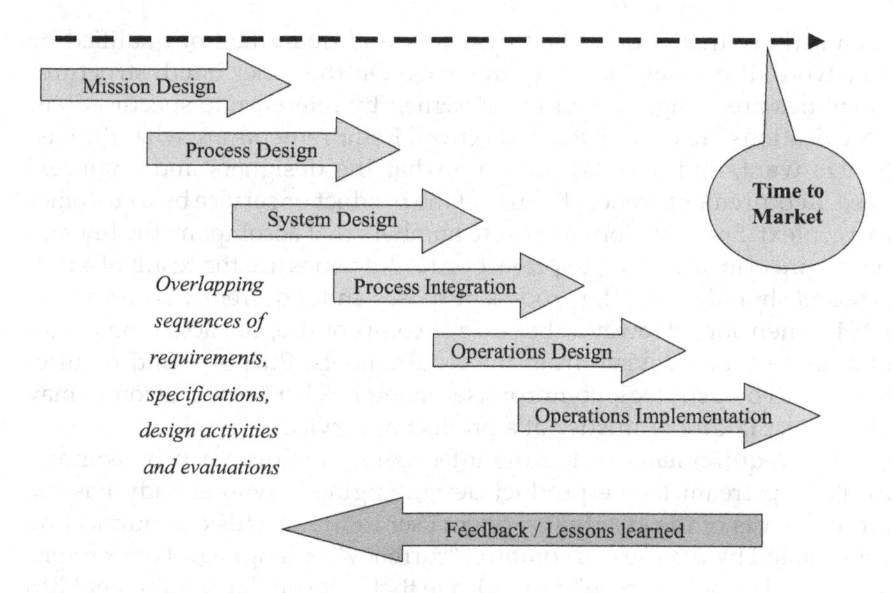


*Figure 4.13* Concurrent engineering.

## Communication in the design process

Communication is an important issue in the design process. A useful language is an adaptation of QFD (Figure 4.14), which has been used in the automobile industry for a couple of decades. It is important to separate the concepts of requirements and specifications. It is also important to distinguish between structures, processes, and outcomes (Figure 4.5). A process

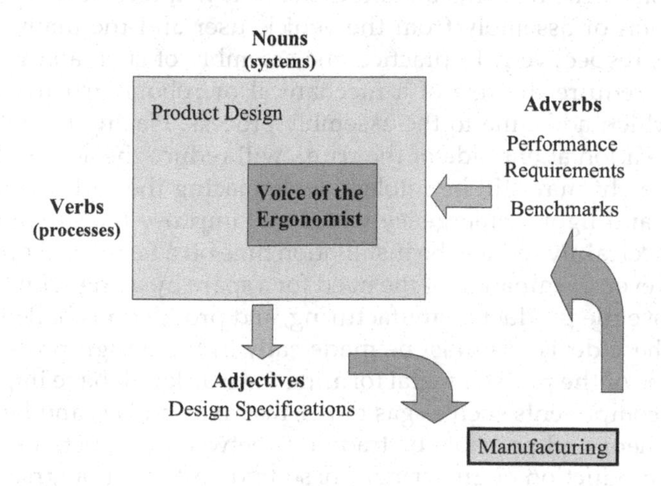


*Figure 4.14* An adaptation of QFD.

is an activity that is described by a verb and quantified or qualified by an adverb. Processes have requirements. On the other hand, structures or entities are things that can be designed by reference to specifications. Specifications are quantitative adjectives. Requirements are what the customers want, and specifications are what the designers and engineers need. Requirements reflect the use of the product or service by a customer in a context. Specifications are those numbers that accompany the lines on the engineering drawing or project plan. Outcomes are the result of a process and should reflect the process purposes and requirements on various E4S4 dimensions. However, because of compromise, the actual outcomes of a process may deviate from the requirements. Purposes and requirements will be a strategic compromise among E4S4 criteria. Outcomes may also reflect failure or misuse of a product or service.

The requirements of the manufacturing employee can be communicated upstream to the product design engineer, who already has the requirements of the eventual product user in mind. This communication is facilitated by the use of a common "currency" or language. For example, a designation of "A/green" can indicate that a particular requirement has been met by a design specification. A "B/yellow" will indicate that a particular design dimension may interact with another criterion to create an intolerable situation. A "C/orange" will suggest that the condition will certainly interact with other criteria. Finally, a "D/red" will indicate that the specified design of an individual dimension is intolerable no matter what the other conditions are.

The case of a spare wheel in a large car will be used to illustrate this process. From both the product use and manufacturing points of view, a large wheel placed horizontally in a well in the bottom of the trunk will be both difficult to install and extract – it will receive a "D" for ease of extraction or assembly from the vehicle user and the manufacturing employee, respectively. In practice, the assembly of such a large component may require the use of a mechanical or robotic articulating arm, both of which add time to the assembly process. Placing the wheel in a vertical location at one side of the trunk will reduce the access challenge but the weight may still be intolerable. Replacing the full-size wheel by a smaller and lighter emergency wheel will improve the situation somewhat and certainly reduce the installation time on a fast-paced production line. However, elimination of the need for a spare by using "run flat" tires will remove the product, manufacturing, and production challenges altogether. These decisions must be made early in the design process as the basic shape of the pressed metal forming the trunk will have implications for other components such as gas tanks, rear axle, brakes, and lighting. In the end, there will probably be trade-offs between product, manufacturing, and production engineering. These trade-offs will be greatly facilitated by a simple metric such as the "A, B, C, D" one described above.

The translation between the operations and manufacturing requirements designations and the weight, location, frequency specifications may require an in depth analysis using an ergonomics tool such as the NIOSH lift Equation. This complex tool assesses adjectives – object weight, vertical and horizontal location, grasping interface and the temporal dimensions of frequency and shift length. The composite NIOSH Lifting Index 1, 2, 3, etc. will translate easily into the A, B, C, and D "adverbs" of the assembly requirements.

An assembly worksheet can be developed that addresses the postural (location), force, and frequency aspects of all assembly tasks. A similar worksheet using the E4S4 or 6Us and 2Ms criteria could be developed for the vehicle owner and the maintainer. Eventually, a composite picture of the whole product life cycle can be used to show the clusters of satisfactory and unsatisfactory features from many viewpoints. The use of color is a very effective way of showing a "forest full of trees."

## Conclusions

The ergonomist has useful contributions to offer at all stages of a product life cycle by representing the "voice of the customers" – user, maintainer, manufacturing employee – in the design process. The ergonomist is effectively and efficiently supported by a tool kit that not only includes in depth analytic tools but also simple communication tools such as E4S4, 6Us and 2Ms, and QFD.

## References

Akao, Y. (1990). *Quality Function Deployment: Integrating Customer Requirements into Product Design*. Productivity Press, New York.

Donabedian, A. (1988). The quality of care, *JAMA* 260 (12), 1743–1748.

Peacock, B. (2014). Ergonomics of design. In A.B. Badiru (Ed.), *Handbook of Industrial and Systems Engineering*. CRC Press, Boca Raton, FL.

Peacock, B. and Karwowski, W. (1993). *Automotive Ergonomics*. Taylor & Francis Group, Boca Raton, FL.

Peacock, B. and Resnick, M. (2010). The 6Us, *Applied Ergonomics*, 41 (1) 130–137.

# chapter five

# Systems theory for systems modeling[1]

## Introduction

Systems theory is a term frequently mentioned in the systems literature. As currently used, systems theory is lacking a universally agreed upon definition. Examples of multiple definitions are provided in Table 5.1. Two of the definitions in Table 5.1 refer to General Systems Theory, a concept espoused by Ludwig von Bertalanffy, Kenneth Boulding, Anatol Rapoport, and Ralph Gerard in the original 1954 bylaws for the foundation of the Society for General Systems Research (SGSR). The aims of General Systems Theory (GST), as stated in the SGSR bylaws, were [Hammond, 2002: 435–436]:

1. To investigate the isomorphy of concepts, laws, and models from various fields, and to help in useful transfers from one field to another
2. To encourage development of adequate theoretical models in fields which lack them
3. To minimize the duplication of theoretical effort in different fields
4. To promote the unity of science through improving communications among specialists

Peter Checkland [1993: 93] remarked that "the general theory envisaged by the founders has certainly not emerged, and GST itself has recently been subject to sharp attacks by both Berlinski (1976) and Lilienfield (1978)." We believe that this is because GST [Bertalanffy, 1968] did not provide either a construct for systems theory or the supporting axioms and propositions required to fully articulate and operationalize a theory.

In order to improve the depth of understanding for systems practitioners using the term *systems theory*, we believe that a more unifying definition and supporting construct need to be articulated. Although there is not a generally accepted canon of general theory that applies to systems, we believe that there are a number of individual systems propositions

[1] Adapted and reprinted from Adams, Kevin MacGregor, Patrick T. Hester, Joseph M. Bradley, Thomas J. Meyers, and Charles B. Keating (2014), "Systems Theory as the Foundation for Understanding Systems," ***Systems Engineering***, Vol. 17, No. 1, pp. 112–123, 2014.

**Table 5.1** Definitions for systems theory

| Definition | Author and year |
|---|---|
| *The formal correspondence of general principles, irrespective of the kinds of relations or forces between the components, lead to the conception of a 'General Systems' Theory as a new scientific doctrine, concerned with the principles which apply to systems in general.* | Bertalanffy (1950b) |
| *General systems theory is the skeleton of science in the sense that it aims to provide a framework or structure of systems on which to hang flesh and blood of particular disciplines and particular subject mat lets in an orderly and coherent corpus of knowledge.* | Boulding (1956) |
| *A new way of looking at the world in which individual phenomena are viewed as interrelated rather than isolated, and complexity has become a subject of interest.* | Klir (1972) |
| *General Systems Theory and the Systems Approach grapple with the issue of 'simplicity' and 'complexity' by which the relationships among systems and subsystems are decided. The problems of 'optimization' and 'suboptimization' are central to explaining the fruitless efforts of systems designers who reach for the 'summum bonum' while settling for a 'second best'.* | van gigch (1974) |

that are relevant to a common practical perspective for systems theory. We therefore propose a formal definition and supporting construct for systems theory.

We propose that systems theory is a unified group of specific propositions which are brought together to aid in understanding systems, thereby invoking improved explanatory power and interpretation with major implications for systems practitioners. It is precisely this group of propositions that enables thinking and action with respect to systems. However, there is no one specialized field of endeavor titled *systems* from which systems theory may be derived. Rather, the propositions available for inclusion into a theory of systems come from a variety of disciplines, thereby making its underlying theoretical basis inherently multidisciplinary. This paper will (1) discuss the functional fields of science in which systems theory can be grounded, (2) provide a definition, construct, and proposed taxonomy of axioms (an axiom set) for systems theory and its associated supporting propositions, derived from the fields of science, and (3) conclude by providing an introductory view of the multidisciplinary breadth represented by systems theory.

## *Individual fields of science*

We propose that science has a hierarchical structure for knowledge contributions as shown in Table 5.2. The Organization for Economic Co-operation

*Table 5.2* Structure for knowledge contributions

| Level | Basic description |
|---|---|
| Philosophical | The emerging system of beliefs providing grounding for theoretical development |
| Theoretical | Research focused on explaining phenomena related to scientific underpinnings and development of explanatory models and testable conceptual frameworks |
| Methodological and axiomatic | Investigation into the emerging propositions, concepts, and laws that define the field and provide high level guidance for design and analysis |
| Technique | Specific models, technologies, standards, and tools for implementation |

and Development (OECD) has provided an internationally accepted classification for the fields of science [OECD, 2007]. This classification includes six major sectors and 42 individual fields of science. The major sectors and individual fields of science are described in Table 5.3. The 42 individual fields of science in Table 5.3 serve as the source for the propositions that are brought together to form a construct for systems theory.

These structural elements constitute the major contributions on which each scientific field's body of knowledge is founded. We display this concept by using a series of concentric rings where the level of knowledge contribution (Table 5.2) radiates from the center and each of the 42 specific fields of science (Table 5.3) is a sector on the circle. Figure 5.1 is a simplified diagram of how we can account for the knowledge from within a functional field of science.

## Systems theory

We believe that the underlying theoretical basis developed in this paper will provide an appropriate foundation for understanding systems. Understanding the axioms and propositions that underlie all systems is mandatory for developing a universally accepted construct for systems theory. The paragraphs that follow will describe our notion of theory, propose a group of constituent propositions, construct a set of proposed axioms, and provide a construct for systems theory.

### Introduction to theory

Theory is defined in a variety of ways. Table 5.4 is a collection of definitions for theory and the key elements associated with each. From these definitions it should be clear that a theory does not have a single proposition that defines it, but is a population of propositions (i.e., arguments,

*Table 5.3* Major and Individual Fields of Science

| Major Fields of Science | Natural Science | Engineering and Technology | Medical and Health Sciences | Agricultural Sciences | Social Sciences | Humanities |
|---|---|---|---|---|---|---|
| **Individual Fields of Science** | 1. Mathematics | 8. Civil engineering | 19. Basic medicine | 24. Agriculture, forestry, and fisheries | 29. Psychology | 38. History and archaeology |
| | 2. Computer and information sciences | 9. Electrical engineering, electronic engineering, information engineering | 20. Clinical medicine | 25. Animal and dairy science | 30. Economics and business | 39. Languages and literature |
| | 3. Physical sciences | 10. Mechanical engineering | 21. Health sciences | 26. Veterinary science | 31. Educational sciences | 40. Philosophy, ethics and religion |
| | 4. Chemical sciences | 11. Chemical engineering | 22. Health biotechnology | 27. Agricultural biotechnology | 32. Sociology | 41. Art (arts, history of arts, performing arts, music) |
| | 5. Earth and related environmental sciences | 12. Materials engineering | 23. Other medical sciences | 28. Other agricultural sciences | 33. Law | 42. Other humanities |
| | 6. Biological sciences | 13. Medical engineering | | | 34. Political Science | |

(*Continued*)

***Table 5.3 (Continued)*** Major and Individual Fields of Science

| Major Fields of Science | Natural Science | Engineering and Technology | Medical and Health Sciences | Agricultural Sciences | Social Sciences | Humanities |
|---|---|---|---|---|---|---|
| | 7. Other natural sciences | 14. Environmental engineering<br>15. Environmental biotechnology<br>16. Industrial biotechnology<br>17. Nano-technology<br>18. Other engineering and technologies | | | 35. Social and economic geography<br>36. Media and communications<br>37. Other Social Sciences | |

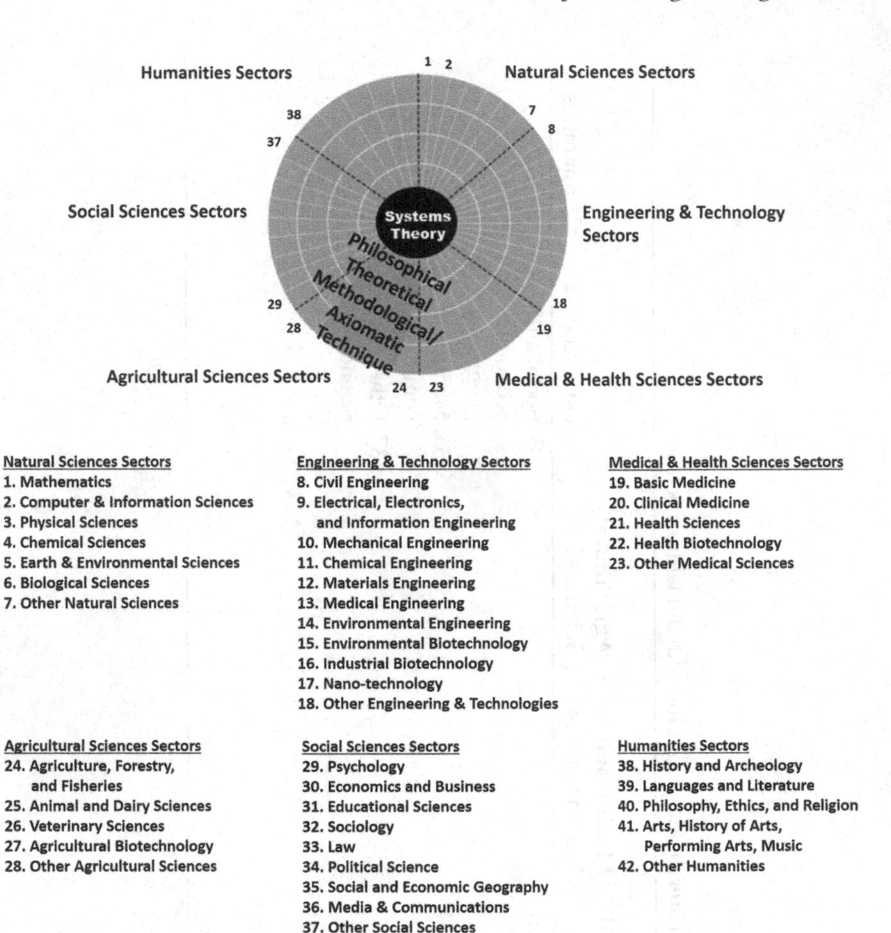


***Figure 5.1*** Depiction of knowledge and the fields of science.

hypotheses, predictions, explanations, and inferences) that provide a skeletal structure for explanation of real-world phenomena. Drawing on the literature, we define theory as follows:

> A unified system of propositions made with the aim of achieving some form of understanding that provides an explanatory power and predictive ability.

The relationship between theory and its propositions is not a direct relationship. It is indirect, through the intermediary of the axioms, where the links in the theory represent the correspondence through similarity to the empirical, real-world system. Figure 5.2 depicts these relationships.

*Table 5.4* Definitions for Theory

| Definition | Key Elements |
|---|---|
| A scientific theory is an attempt to bind together in a systematic fashion the knowledge that one has of some particular aspect of the world of experience. The aim is to achieve some form of understanding, where this is usually cashed out as an explanatory power and predictive fertility (Honderich, 2005, p. 914). | • Bind together in a systematic fashion<br>• Explanatory power and predictive fertility |
| A unified system of laws or hypotheses, with explanatory force (Proudfoot and Lacev, 2010). | • Unified system |
| We understand a theory as comprising two elements: (1) a population of models, and (2) various hypotheses linking those models with systems in the real world (Giere, 1988). | • Population of models<br>• Linked to the real world through hypotheses |
| An abstract calculus is the logical skeleton of the explanatory system, and implicitly defines the basic notions of the system. A set of rules that assign an empirical content to the abstract calculus by relating it to the concrete materials of observation and experiment. An interpretation or model for the abstract calculus, which supplies some flesh for the skeletal structure in terms of more or less familiar conceptual or visualizable materials (Nagel, 1961). | • Logical skeleton of the explanatory system<br>• Set of rules<br>• Model for the abstract calculus, which supplies some flesh for the skeletal structure |
| A coherent set of principles or statements that explains a large set of observations or findings (Angier, 2007). | • Set of propositions (Angier's principles)<br>• Explains a large set of observations |

Our notion of theory is a population of propositions that "... explains a [real system in terms of a] large set of observations or findings. Those constituent findings are the product of scientific research and experimentation, those findings, in other words, already have been verified, often many times over, and are as close to being 'facts' as science cares to characterize them" [Angier, 2007: 154]. Our representation of theory subscribes to the model espoused by Giere [1988:87] where "rather than regarding the axioms and theorems as empirical claims, treat them all merely as definitions." In this case, our model of systems theory is defined by its set of axioms and supporting propositions.

The following section will use the axiomatic method [Audi, 1999, p. 65] to articulate the accepted propositions and concepts from the 42 fields of science discussed previously in order to increase certainty in the propositions and clarity in the concepts we propose as *systems theory.*

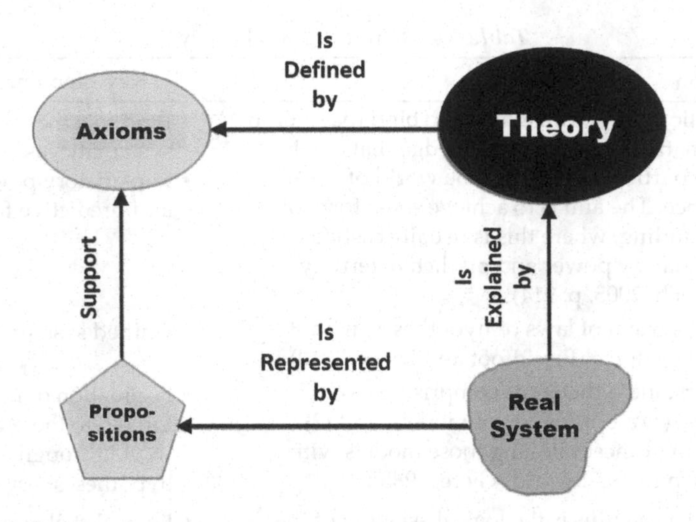


*Figure 5.2* Relationship between theory, propositions, axioms, and real system.

## *Systems propositions*

This section addresses a proposed group of constituent propositions that we have encountered in our work with systems. Each of the propositions has an empirical basis in one of the 42 individual fields of science in Table 5.3. While likely incomplete, this set of propositions provides a representation of real-world systems encountered during our work with systems problems. Each underlying proposition, its primary proponent in the literature, and brief descriptions are presented in Table 5.5a and Table 5.5b.

## *Axioms of systems theory*

This section addresses a proposed set of axioms and their constituent propositions that we termed *systems theory*. The 30 propositions presented supported inductive development of the axioms. Using the axiomatic method [Audi, 1999], the propositions were reorganized into seven axioms as follows:

- The Centrality Axiom states that *central to all systems are two pairs of propositions: emergence and hierarchy, and communication and control.* The centrality axiom's propositions describe the system by focusing on (1) a system's hierarchy and its demarcation of levels based on emergence and (2) systems control which requires feedback of operational properties through communication of information.

***Table 5.5a*** Alphabetical Listing of Systems Propositions

| Proposition and Primary Proponent | Brief Description of the Systems Proposition |
|---|---|
| Circular causality (Korzybski, 1994) | An effect becomes a causative factor for future effects, influencing them in a manner particularly subtle, variable, flexible, and of an endless number of possibilities. |
| Communication (Shannon, 1948a, 1948b) | In communication, the amount of information is defined, in the simplest cases, to be measured by the logarithm of the number of available choices. Because most choices are binary, the unit of information is the *bit*, or binary digit. |
| Complementarity (Bohr, 1928). | Two different perspectives or models about a system will reveal truths regarding the system that are neither entirely independent nor entirely compatible. |
| Control (Checkland,1993) | The process by means of which a whole entity retains its identity and/or performance under changing circumstances. |
| Darkness (Cilliers, 1998) | Each element in the system is ignorant of the behavior of the system as a whole, it responds only to information that is available to it locally. This point is vitally important. If each element 'knew' what was happening to the system as a whole, all of the complexity would have to be present in that element (Cilliers, 1998). |
| Dynamic equilibrium (D'Alembert, 1743) | For a system to be in a state of equilibrium, all subsystems must be in equilibrium. All subsystems being in a state of equilibrium, the system must be in equilibrium. |
| Emergence (Aristotle, 2002) | Whole entities exhibit properties which are meaningful only when attributed to the whole, not its parts - e.g., the smell of ammonia. Every model of systems exhibits properties as a whole entity which derive from it component activities and their structure, but cannot be reduced to them (Checkland, 1993). |
| Equifinality (Bertalanffy, 1950a) | If a steady state is reached in an open system, it is independent of the initial conditions, and determined only by the system parameters, i.e., rates of reaction and transport. |
| Feedback (Wiener, 1948) | All purposeful behavior may be considered to require negative feed-back. If a goal is to be attained, some signals from the goal are necessary at some time to direct the behavior. |
| Hierarchy (Pattee, 1973) | Entities meaningfully treated a wholes are built up of smaller entities which are themselves wholes … and so on. In a hierarchy, emergent properties denote the levels (Checkland, 1993). |

(*Continued*)

***Table 5.5a (Continued)*** Alphabetical Listing of Systems Propositions

| Proposition and Primary Proponent | Brief Description of the Systems Proposition |
|---|---|
| Holism (Smuts, 1926) | The whole is not something additional to the parts: it is the parts in a definite structural arrangement and with mutual activities that constitute the whole. The structure and the activities differ in character according to the stage of development of the whole; but the whole is just this specific structure of parts with their appropriate activities and functions (Smuts, 1926). |
| Homeorhesis (Waddington, 1957, 1968) | The concept encompassing dynamical systems which return to a trajectory, as opposed to systems which return to a particular state, which is termed homeostasis. |
| Homeostasis (Cannon, 1929) | The property of an open system to regulate its internal environment so as to maintain a stable condition, by means of multiple dynamic equilibrium adjustments controlled by interrelated regulation mechanisms. |
| Information Redundancy (Shannon and Weaver, 1949) | The number of bits used to transmit a message minus the number of bits of actual information in the message. |
| Minimum Critical Specification (Cherns, 1976, 1987) | This principle has two aspects, negative and positive. The negative simply states that no more should be specified than is absolutely essential; the positive requires that we identify what is essential. |

***Table 5.5b*** Continuation of Alphabetical Listing of Systems Propositions

| Proposition and Primary Proponent | Brief Description of the Systems Proposition |
|---|---|
| Multifinality (Buckley, 1967) | Radically different end states are possible from the same initial conditions. |
| Pareto (Pareto, 1897) | Eighty percent of the objectives or outcomes are achieved with 20% of the means. |
| Purposive behavior (Rosenblueth, Wiener, & Bigelow, 1943) | Purposeful behavior is meant to denote that the act or behavior may be interpreted as directed to the attainment of a goal-i.e., to a final condition in which the behaving object reaches a definite correlation in time or in space with respect to another object or event. |
| Recursion (Beer, 1979) | The fundamental laws governing the processes at one level are also present at the next higher level. |

*(Continued)*

***Table 5.5b (Continued)*** Continuation of Alphabetical Listing of Systems Propositions

| Proposition and Primary Proponent | Brief Description of the Systems Proposition |
|---|---|
| Redundancy (Pahl et al., 2011) | Means of increasing both the safety and reliability of systems by providing superfluous or excess resources. |
| Redundancy of potential command (McCulloch, 1959) | Effective action is achieved by an adequate concatenation of information. In other words, power resides where information resides. |
| Relaxation time (Holling, 1996) | Stability near an equilibrium state, where resistance to disturbance and speed of return to the equilibrium are used to measure the property. The system's equilibrium state is shorter than the mean time between disturbances. |
| Requisite hierarchy (Aulin-Ahmavaara, 1979) | The weaker in average are the regulatory abilities and the larger the uncertainties of available regulators, the more hierarchy is needed in the organization of regulation and control to attain the same result, if possible at all. |
| Requisite parsimony (Miller, 1956) | Human short-term memory is incapable of recalling more than seven plus or minus two items (Simon, 1974). |
| Requisite saliency (Boulding, 1966) | The factors that will be considered in a system design are seldom of equal importance. Instead, there is an underlying logic awaiting discovery in each system design that will reveal the saliency of these factors. |
| Requisite variety (Ashby, 1956) | Control can be obtained only if the variety of the controller is at least as great as the variety of the situation to be controlled. |
| Satisficing (Simon, 1955, 1956) | The decision making process whereby one chooses an option that is, while perhaps not the best, good enough. |
| Self-organization (Ashby, 1947) | The spontaneous emergence of order out of the local interactions between initially independent components. |
| Suboptimization (Hitch, 1953) | If each subsystem, regarded separately, is made to operate with maximum efficiency, the system as a whole will not operate with utmost efficiency. |
| Viability (Beer, 1979) | A function of balance must be maintained along two dimensions: (1) autonomy of subsystem versus integration and (2) stability versus adaptation. |

- The Contextual Axiom states that *system meaning is informed by the circumstances and factors that surround the system*. The contextual axiom's propositions are those which bound the system by providing guidance that enables an investigator to understand the set of external circumstances or factors that enable or constrain a particular system.
- The Goal Axiom states that *systems achieve specific goals through purposeful behavior using pathways and means*. The goal axiom's propositions address the pathways and means for implementing systems that are capable of achieving a specific purpose.
- The Operational Axiom states that *systems must be addressed in situ, where the system is exhibiting purposeful behavior*. The operational axiom's propositions provide guidance to those that must address the system in situ, where the system is functioning to produce behavior and performance.
- The Viability Axiom states that *key parameters in a system must be controlled to ensure continued existence*. The viability axiom addresses how to design a system so that changes in the operational environment may be detected and affected to ensure continued existence.
- The Design Axiom states that *system design is a purposeful imbalance of resources and relationships*. Resources and relationships are never in balance because there are never sufficient resources to satisfy all of the relationships in a systems design. The design axiom provides guidance on how a system is planned, instantiated, and evolved in a purposive manner.
- The Information Axiom states that *systems create, possess, transfer, and modify information*. The information axiom provides understanding of how information affects systems.

The specific axiom and its supporting propositions are presented in Table 5.6. It is important to note that neither propositions nor their associated axioms are independent of one another.

## *Construct for systems theory*

Systems theory provides explanations for real-world systems. These explanations increase our understanding and provide improved levels of explanatory power and predictive ability for the real-world systems we encounter. Our view of systems theory is a model of linked axioms (composed of constituent propositions) that are represented through similarity to the real system [Giere, 1988]. Figure 5.3 is a construct of the axioms of systems theory. The axioms presented are called the "theorems of the system or theory" [Honderich, 2005] and are the set of axioms, presumed

***Table 5.6*** Axioms for Systems Theory

| Axiom | Proposition and Primary Proponent |
|---|---|
| Centrality | Communication (Shannon, 1948a, 1948b) |
| | Control (Checkland, 1993) |
| | Emergence (Aristotle, 2002) |
| | Hierarchy (Pattee, 1973) |
| Contextual | Complementarity (Bohr, 1928) |
| | Darkness (Cilliers, 1998) |
| | Holism (Smuts, 1926) |
| Design | Minimum Critical Specification (Cherns, 1976, 1987) |
| | Pareto (Pareto, 1897) |
| | Requisite Parsimony (Miller, 1956) |
| | Requisite Saliency (Boulding, 1966) |
| Goal | Equifinality (Bertalanffy, 1950a) |
| | Multifinality (Buckley, 1967) |
| | Purposive behavior (Rosenblueth et al., 1943) |
| | Satisficing (Simon, 1955, 1956) |
| | Viability (Beer, 1979) |
| Information | Redundancy of Potential Command (McCulloch, 1959) |
| | Information Redundancy (Shannon & Weaver, 1949) |
| Operational | Dynamic equilibrium (D'Alembert, 1743) |
| | Homeorhesis (Waddington, 1957, 1968) |
| | Homeostasis (Cannon, 1929) |
| | Redundancy (Pahl et al., 2011) |
| | Relaxation Time (Holling, 1996) |
| | Self-organization (Ashby, 1947) |
| | Suboptimization (Hitch, 1953) |
| Viability | Circular causality (Korzybski, 1994) |
| | Feedback (Wiener, 1948) |
| | Recursion (Beer, 1979) |
| | Requisite hierarchy (Aulin-Ahmavaara, 1979) |
| | Requisite variety (Ashby, 1956) |

true by systems theory, from which all other propositions in systems theory may be induced.

Systems theory is the unified group of propositions, linked with the aim of achieving understanding of systems. Systems theory, as proposed in this paper, will permit systems practitioners to invoke improved explanatory power and predictive ability. It is precisely this group of propositions that enables thinking, decision, action, and interpretation with respect to systems.

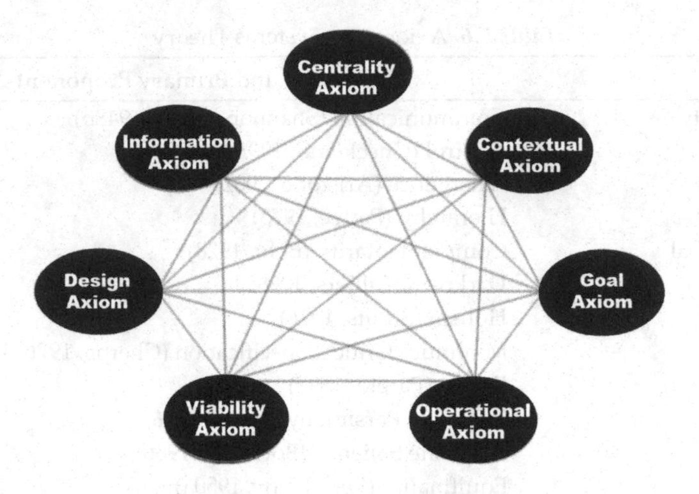


*Figure 5.3* Axioms of systems theory.

The axiom set in Figure 5.3 may be considered a construct of a system, where a construct is defined as a characteristic that cannot be directly observed and so can only be measured indirectly [Bernard, 2002; Gliner and Morgan, 2000; Leedy and Ormrod, 2001; Orcher, 2005] and a system is defined as "...a set of interrelated components working together toward some common objective or purpose" [Blanchard and Fabrycky, 2006: 2]. Thus, a system may be identified as such if it exhibits and can be understood within this set of axioms. Conversely, any entity that exhibits these seven axioms is, by definition, a system. Thus, given its testable nature, this construct can be evaluated with respect to systems under consideration in order to determine its generalizability. Further, given the multidisciplinary nature of its foundational axioms and the multidisciplinary nature under which the construct was formed, there are numerous implications for multidisciplinary application of such a construct.

## *Multidisciplinary implications of systems theory*

We have presented a construct for systems theory, proposed a set of seven axioms and group of supporting propositions from the 42 fields of science. Our construct for systems theory is the unified group of propositions, linked by an axiom set that aims to achieve understanding of systems that provides improved explanatory power and predictive ability. It is precisely this group of propositions that enables thinking, decision, action, and interpretation with respect to systems.

We believe that systems theory is the foundation for understanding multidisciplinary systems. Practitioners can benefit from the application of systems theory as a lens when viewing multidisciplinary systems and their related problems. Systems theory and the associated language of systems are important enabling concepts for systems practitioners. The set of seven framework axioms and associated group of propositions that we designate as systems theory allow systems practitioners to ground their observations to a rigorously developed systems-based foundation.

Behaviors expected from systems should be described by the axioms proposed in this paper. For example, any system should exhibit sub-optimization. For a system as complex as a Boeing 747, this means trade-offs between increased cargo carrying capacity and maximum airspeed, whereas a simpler system such as a laptop computer may require that the heating system be suboptimal (i.e., larger than ideal) in order to support a faster processing chip. While this simply illustrates the use of one of the propositions described herein, each axiom and its associated propositions provides insight into the behavior of the system. Understanding of the proposed construct of systems theory affords systems practitioners greater overall system understanding.

Finally, the propositions from the seven axioms, described briefly in Table 5.5, can be superimposed on the Depiction of Knowledge and the Fields of Science presented in Figure 5.1. Figure 5.4 presents systems theory as the intersection of a number of well-defined multidisciplinary propositions by distinguished authors from the 42 fields of science.

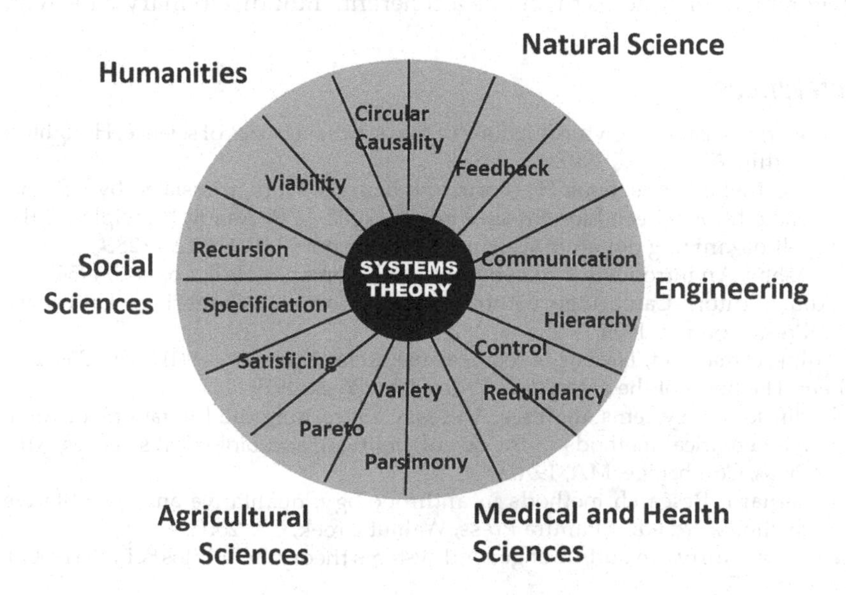


*Figure 5.4* Systems theory and the major fields of science.

It is clear from viewing Figure 5.4 that systems theory and its theoretical foundation are inherently multidisciplinary. Contributions to our perspective of systems theory are incorporated from each of the major fields of science with the exception of agricultural sciences (most probably due to the *darkness* proposition). This multidisciplinary construct ensures widespread applicability of this theory and removes barriers that traditional engineering-centric views of systems place on approaches to problem solving. The lack of a prescription regarding domain applicability further ensures that systems theory is multidisciplinary in both its theoretical foundations and application.

## Conclusion

We have proposed systems theory as a unified group of specific propositions which are brought together by way of an axiom set to form the construct of a system. This construct affords systems practitioners and theoreticians with a prescriptive set of axioms by which the system operation can be understood; conversely, any entities identified as a system may be characterized by this set of axioms. Given its multidisciplinary theoretical foundation and multidisciplinary framework, systems theory, as developed in this paper, is posited as a general approach to aid in understanding system behavior. This formulation is in its embryonic stages and would be well served from feedback and challenge from systems practitioners to test this proposed construct and encourage future development of systems theory as a coherent, multidisciplinary endeavor.

## References

N. Angier, The canon: A whirligig tour of the beautiful basics of science, Houghton Mifflin, New York, 2007.

Aristotle, Metaphysics, Book H—Form and being at work, translated by J. Sachs, 2nd edition, Green Lion Press, Sante Fe, 2002. W.R. Ashby, Principles of the self-organizing dynamic system, J Gen Psychol 37 (1947), 125–128.

W.R. Ashby, An introduction to cybernetics, Chapman & Hall, London, 1956.

R. Audi (Editor), Cambridge dictionary of philosophy, Cambridge University Press, London, 1999.

A. Aulin-Ahmavaara, The law of requisite hierarchy, Kybernetes 8(4) (1979), 259–266.

S. Beer, The heart of the enterprise, Wiley, New York, 1979.

D. Berlinski, On systems analysis: An essay concerning the limitations of some mathematical methods in the social, political, and biological sciences, MIT Press, Cambridge, MA, 1976.

H.R. Bernard, Research methods in anthropology: qualitative and quantitative methods, 3rd ed., Altamira Press, Walnut Creek, CA, 2002.

L. von Bertalanffy, An outline of general systems theory, Br J Philos Sci 1(2) (1950a), 134–165.

L. von Bertalanffy, The theory of open systems in physics and biology, Science 111(2872) (1950b), 23–29.
L. von Bertalanffy, General system theory: Foundations, development, applications, rev. ed., Braziller, New York, 1968.
B.S. Blanchard and W.J. Fabrycky, Systems engineering and analysis, 4th ed., Prentice–Hall, Upper Saddle River, NJ, 2006.
N. Bohr, The quantum postulate and the recent development of atomic theory, Nature 121(3050) (1928), 580–590.
K. Boulding, General systems theory—The skeleton of science, Manage Sci 2(3) (1956), 197–208.
K. Boulding, The impact of social sciences, Rutgers University Press, New Brunswick, NJ, 1966.
W. Buckley, Sociology and modern systems theory, Prentice-Hall, Englewood Cliffs, NJ, 1967.
W.B. Cannon, Organization for physiological homeostasis, Physiol Rev 9(3) (1929), 399–431.
P.B. Checkland, Systems thinking, systems practice, Wiley, New York, 1993.
A. Cherns, The principles of sociotechnical design, Hum Relat 29(8) (1976), 783–792.
A. Cherns, The principles of sociotechnical design revisited, Hum Relat 40(3) (1987), 153–161.
P. Cilliers, Complexity and postmodernism: Understand complex systems, Routledge, New York, 1998.
J. D'Alembert, Traité de dynamique, David l'Ainé, Paris, 1743. R.N. Giere, Explaining science: A cognitive approach, University of Chicago Press, Chicago, 1988.
J.A. Gliner and G.A. Morgan, Research methods in applied settings: An integrated approach to design and analysis, Erlbaum, Mahwah, NJ, 2000.
D. Hammond, Exploring the genealogy of systems thinking, Syst Res Behav Sci 19(5) (2002), 429–439.
C.J. Hitch, Sub-optimization in operations problems, J Oper Res Soc Am 1(3) (1953), 87–99.
C.S. Holling, "Engineering resilience versus ecological resilience," in P. Schulze (Editor), Engineering within ecological constraints, National Academies Press, Washington, DC, 1996, pp. 31–43.
T. Honderich, The Oxford companion to philosophy, 2nd ed., Oxford University Press, New York, 2005, pp. 1–1056.
G.J. Klir, "Preview: The polyphonic GST," in G.J. Klir (Editor), Trends in general systems theory, Wiley, New York, 1972, pp. 1–16.
A. Korzybski, Science and sanity: An introduction to non-Aristotelian systems and general semantics, Wiley, New York, 1994.
P.D. Leedy and J.E. Ormrod, Practical research planning and design, 9th ed., Pearson Education, Upper Saddle River, NJ, 2001.
R. Lilienfield, The rise of systems theory: An ideological analysis, Wiley, New York, 1978.
W.S. McCulloch, Embodiments of mind, MIT Press, Cambridge, MA, 1959.
G. Miller, The magical number seven, plus or minus two: Some limits on our capability for processing information, Psychol Rev 63(2) (1956), 81–97.
E. Nagel, The structure of science, Harcourt, Brace and Wilson, New York, 1961.

OECD, Revised field of science and technology (FOS) classification in the Frascati manual, Organization for Economic Cooperation and Development, Paris, 2007.

L.T. Orcher, Conducting research: social and behavioral science methods, Pyrczak, Glendale, CA, 2005.

G. Pahl, W. Beitz, J. Feldhusen, and K.-H. Grote, Engineering design: a systematic approach (K. Wallace and L.T.M. Blessing, Trans. 3rd ed.), Springer, Berlin, 2011.

V. Pareto, Cours d'économie politique professé à l'Université de Lausanne, University of Luzerne, Luzerne, 1897.

H.H. Pattee, Hierarchy theory: The challenge of complex systems, Braziller, New York, 1973, pp. 1–156.

M. Proudfoot and A.R. Lacey, The Routledge dictionary of philosophy, 4th ed., Routledge, Abingdon, 2010.

A. Rosenblueth, N. Wiener, and J. Bigelow, Behavior, purpose and telelogy, Philos Sci 10(1) (1943), 18–24.

C.E. Shannon, A mathematical theory of communication, Part 1, Bell Syst Tech J 27(3) (1948a), 379–423.

C.E. Shannon, A mathematical theory of communication, Part 2, Bell Syst Tech J 27(4) (1948b), 623–656.

C.E. Shannon and W. Weaver, The mathematical theory of communication, University of Illinois Press, Champaign, 1949.

H.A. Simon, A behavioral model of rational choice, Q J Econ 69(1) (1955), 99–118.

H.A. Simon, Rational choice and the structure of the environment, Psychol Rev 63(2) (1956), 129–138.

H.A. Simon, How big is a chunk?, Science 183(4124) (1974), 482–488.

J. Smuts, Holism and evolution, Greenwood Press, New York, 1926.

J. van Gigch, Applied general systems theory, 2nd ed., Harper and Row, New York, 1974.

C.H. Waddington, The strategy of genes: A discussion of some aspects of theoretical biology, Allen & Unwin, London, 1957.

C.H. Waddington, Towards a theoretical biology, Nature 218(5141) (1968), 525–527.

N. Wiener, Cybernetics: Or control and communication in the animal and the machine, MIT Press, Cambridge, MA, 1948.

# *chapter six*

# *System-of-systems engineering management for systems modeling*[1]

## *Introduction*

In this chapter, we review the SoS literature to illustrate the need to create an SoSE management framework based on the demands of constant technological progress in a complex dynamic environment. We conclude from this review that the history and evolution of defining SoS has shown that: 1) SoS can be defined by distinguishing characteristics and 2) SoS can be viewed as a network where the "best practices" of network management can be applied to SoSE. We use these two theories as a foundation for our objective to create an effective SoSE management framework. To accomplish this, we utilize modified fault, configuration, accounting, performance, and security (FCAPS) network principles (SoSE management conceptual areas). Furthermore, cited distinguishing characteristics of SoS are also used to present a SoSE management framework. We conclude with a case analysis of this framework using a known and well-documented SoS (i.e., Integrated Deepwater System) to illustrate how to better understand, engineer, and manage within the domain of SoSE.

While the research on system of systems (SoS) has shown significant development in studies and experimental applications on this topic, a review of relevant modern literature reveals that we are still in an embryonic state in terms of identifying an effective methodology to achieve the objectives of system-of-systems engineering (SoSE). We can trace the origin of the concept of systems, and thus SoS, to "the Greek word sustema (that) stood for reunion, conjunction or assembly" [1]. From its origin, we can track the evolution of this term as it has been studied through general systems theory, systemics, and cybernetics. Today, IEEE Standard 1220 defines a System as a: "set or arrangement of elements [people, products (hardware and software) and processes (facilities, equipment, material,

[1] Adapted and reprinted from Gorod, Alex; Brian Sauser, and John Boardman (2008), "System-of-Systems Engineering Management: A Review of Modern History and a Path Forward," IEEE Systems Journal, Vol. 2, No. 4, Dec 2008, pp. 484–499.

and procedures)] that are related and whose behavior satisfies customer/operational needs, and provides for the life cycle sustainment of the products" [2].

Just as the term has evolved, so has our understanding of how to conceptualize and realize (i.e., engineer) systems. As the world experienced major structural and operational changes in production and manufacturing around World War II, there was a significant paradigm shift in dealing with new complexities by introducing new engineering techniques that focused on a complex system rather than separate individual components. This became known as the discipline of systems engineering (SE). Still, a process of rapid global acceleration, especially in the military sector, continued and called for the next level of development in engineering. The objective was to address "shortcomings in the ability to deal with difficulties generated by increasingly complex and interrelated system of systems" [3].

There was a need for a discipline that focused on the engineering of multiple integrated complex systems [3]. Today, we refer to this as SoSE. Unfortunately, we are still attempting to understand its principles, practices, and execution. There is no universally accepted definition of SoSE or SoS [4] despite the fact that there have been multiple attempts to create one. For example, Kotov used the definition, "Systems of systems are large scale concurrent and distributed systems that are comprised of complex systems" [5]; Manthorpe's military-specific definition states, "In relation to joint warfighting, system of systems is concerned with interoperability and synergism of Command, Control, Computers, Communications, and Information (C4I) and Intelligence, Surveillance, and Reconnaissance (ISR) Systems" [6]; and Luskasik's education-specific definition states, "SoSE involves the integration of systems of systems that ultimately contribute to evolution of the social infrastructure" [7].

Alternatively, some researchers have taken a different approach by focusing on a characterization rather than providing an abstract definition of SoS [8]–[11]. This characterization approach provides a more comprehensive and precise taxonomy whereas the definitional approach is limited to an industry-specific context and lacks the flexibility necessary for successful dynamic trans-disciplinary engineering processes [12]. In addition, the use of characteristics enables us to better identify the dynamic nature of various forces within SoS [11].

The goal of SoSE has remained consistent in the literature. It is comprised of the successful engineering of multiple integrated complex systems. Thus, with a foundation in complex systems we have also had to embrace the principles of networks. Shenhar [13] was one of the first to describe SoS as a network of systems functioning together to achieve a common purpose.

Later, others including Maier [14] and Lane and Valerdi [15] identified other universally known network-centric systems as examples of collaborative SoS (i.e., the internet, global communication networks, etc.). Recently, Gorod, et al. [16], suggested extracting the best practices of network management that would enable the development of an effective SoSE management framework.

In this chapter, we will bridge these two perspectives of SoS, characterization and networks, to build a framework for understanding how we can realize and manage SoS. The fundamental value in a framework is being able to correctly associate the definition and arrangement of a system [17]. We will first provide a review of the modern literature on SE and SoSE, so we may understand where we have come and where we can go. Second, we will present a characterization approach to describe SoS that has been built on a review of the SoS literature. Third, we will present our SoSE management framework based on this characterization approach and the principles of network management. Finally, we will apply this framework to an SoS case study to exemplify its realization.

## *State of knowledge and practice*

While the body of knowledge for SE has been well documented by many organizations [e.g., International Council on Systems Engineering (INCOSE), IEEE, Department of Defense], Brill's [18] work summarizes the major advancements in the study of SE that took place from 1950 to 1995. We will refer to Brill's work in order to establish what we know about SE. We will then perform a similar exercise as Brill to examine the literature of significant contributors to the body of knowledge for SoSE. This includes journal publications, conference proceedings, standards, guides, government documents, and different relevant events from 1991 to present. We summarize both bodies of knowledge by comparing the major drives that separate SE from SoSE.

### *A. Systems engineering (SE)*

Figure 6.1 depicts Brill's graphical timeline of the key contributors to highlight milestones in the study of SE and its application [18].

According to Hall [19], the first major contributor to the development of SE was Gilman, who probably made the first attempt to teach SE at the Massachusetts Institute of Technology in 1950.

The second significant contribution came from Goode and Machol. They acknowledged the need for a new method of organization through "the systems design, systems analysis, and systems approach" [20].

In 1957, Engstrom, explained the concept of SE through the use of terms such as evolution and characteristics [21].

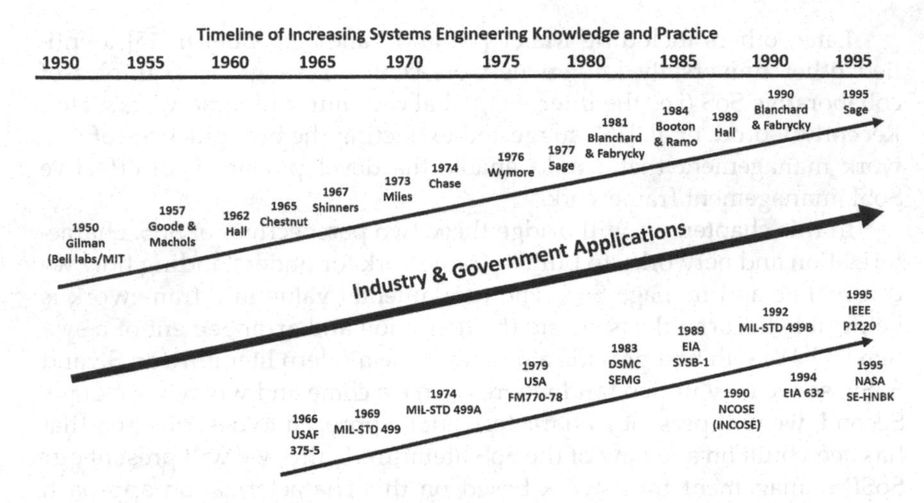


***Figure 6.1*** Increasing SE knowledge and practice [18].

In 1962, Hall introduced a concept of "process of systems engineering" that included three important elements. First, to acquire better knowledge about this complex phenomenon, it was imperative to recognize that systems had to include multi facets in its definition. Second, an engineer had to examine a system from three distinct environmental positions "the physical or technical," "business or economic," and "social." Third, according to the Total Quality Management Principle, to most effectively fulfill customers' objectives, all available knowledge had to be used. This element was further addressed and explained in Hall's Metasystems Methodology [19], [22].

In the 1960's, Shinners and Chestnut, worked to create a methodology to solve any system-oriented problems through first understanding them. In 1967, Shinners proposed to use seven general procedures to generate a feedback process in engineering a large complex system [23]. Chestnut similarly explored an option of the feedback process to best determine a manner to meet the customers' objectives. He suggested several questions in addressing the ability to formulate and solve the problem [24].

In 1973, Miles worked on editing ten lectures by well-recognized members of scientific community on the topic of "Systems Concepts for the Private and Public Sectors" [25]. He articulated a six step approach to meet specific systems engineering goals. He suggested that instead of focusing only on the analysis and design of distinct components, it would be more useful to concentrate on the problem "in its entirety" [25].

In 1974, Chase correctly pointed out that the inadequate state of language development prevented effective communication on topics related to the systems concepts, and there was a lot of work that needed to be done to remedy the situation [26].

In 1976, Wymore put "interdisciplinary team using uniform and standard systems engineering methodology" as a core component in addressing the problem of the "design and analysis of large scale, complex/machine systems" [18], [27]. This methodology primarily focused on the effective communication manner of the interdisciplinary team concept. It included "modeling human behavior, dealing with complexity and largeness-of-scale," and managing dynamic technology [18], [27].

In 1981, Blanchard and Fabrycky introduced the concept of "system-life-cycle-engineering" similar to the studies done by Hall centered on such notions as problem detection and definition; planning and designing of a system; and implementation and obsolescence [19]. They emphasized the need for systems engineers to include all aspects of the system in the proper application of the "system-life-cycle" concept [28].

In 1995, Sage was the first author to suggest that "systems engineering is the management technology that controls a total lifecycle process, which involves and which results in the definition, development, and deployment of a system that is of high quality, trustworthy, and cost effective in meeting user needs" [29]. Sage introduced the key definitions of the systems engineering concept through structure, function, and purpose [29]. Also, at the same time, there were several noteworthy organizations that provided significant contributions to the studies of systems engineering. They published handbooks, standards, and guides related to the advancements in the field.

In 1966, the United States Air Force (USAF) was the first organization to publish a handbook describing a systems engineering process [30]. In 1969, it was replaced with MIL-STD-499 [31].

In 1974, the new MIL-STD-499A introduced the Systems Engineering Management Plan (SEMP) [32]. In 1979, the U.S. Army published guidelines for implementing and managing a systems engineering process [33].

In 1983, The Defense Systems Management College introduced the first edition of the Systems Engineering Management Guide (SEMG) that became popular within the defense industry as a hands-on tool for systems engineering in the military field [34]. In 1989, the Electronic Industries Association (EIA) issued a report to designate systems engineering as "a central process" in meeting "user operational requirements for system/equipment design" [18].

In 1990, the first professional organization was established to study the field of systems engineering. It became known as the National Council on Systems Engineering (NCOSE). In 1992, the U.S. Air Force released an updated version of the first handbook MIL-STD-499A that contained a more comprehensive approach to systems engineering. It was called MIL-STD-499B [35]. In 1994, the EIA, with the assistance of other professional associations, published an Interim Standard 632 that reflected continued efforts to create a defined management process for engineering systems [36]. Finally, in 1995, both IEEE and NASA published their respective reports on systems engineering. The IEEE's PI220 and NASA's SE-HNBK, similar to EIA's Interim Standard 632, focused on the creation of the management process for engineering systems and were generally consistent with the process described by Goode and Machol and Shinners [18], [20], and [23].

In the last 50 years we have seen tremendous progress in our ability to understand, design, develop, implement, and manage single systems. This is an outcome of the research done in the field of SE. Researchers and practitioners from many different disciplines and fields have contributed. As we face a new phenomenon of SoSE the focus has changed. Instead of single systems we now have to cope with multiple integrated complex systems [3]. However, the basic principals of SE can be applied to SoSE. Therefore, it is imperative for us to use SE as a foundation for the research in the field of SoSE.

## B. *System of Systems Engineering (SoSE)*

The initial mention of the SoS can be traced to Boulding [37], Jackson and Keys [38], Ackoff [39], and Jacob [40]. Boulding imagined SoS as a "gestalt" in theoretical construction creating a "spectrum of theories" greater than the sum of its parts. Jackson and Keys suggested using the "SoS methodologies" as interrelationship between different systems-based problem-solving methodologies in the field of operation research. Ackoff considered SoS as a "unified or integrated set" of systems concepts. Jacob stated that a SoS is "every object that biology studies." It was not until 1989, with the Strategic Defense Initiative, that we find the first use of the term "system-of-systems" to describe an engineered technology system [41].

The transition to the accepted modern term SoS is reflected in Figure 6.2 and is introduced in the works of Eisner et al. [42], [43] and Shenhar [13]. Eisner et al. defined SoS as: "A set of several independently acquired systems, each under a nominal systems engineering process; these systems are interdependent and form in their combined operation a multifunctional solution to an overall coherent mission. The optimization of each system does not guarantee the optimization of the overall system of systems" [42].

Shenhar used the term "array" to describe SoS as: "A large widespread collection or network of systems functioning together to achieve a common purpose" [13]. In his more recent works he now describes an "array" as a "system of systems" [44].

In 1995, Holland proposed to study SoS as an artificial complex adaptive system that persistently changes through self-organization with the

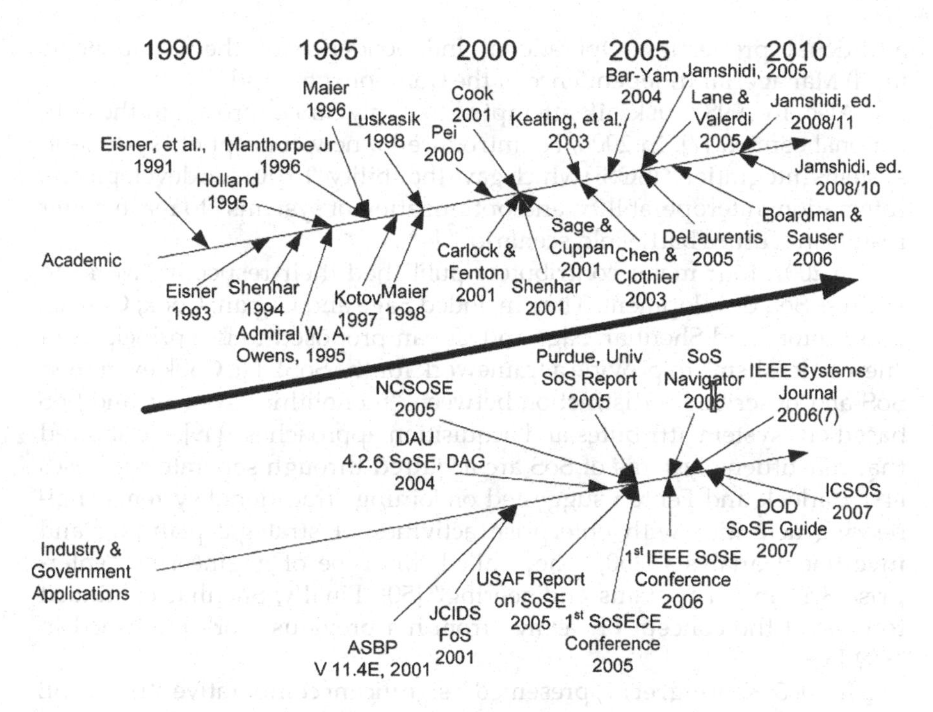


*Figure 6.2* Modern history of SoS.

assistance of local governing rules to adapt to increasing complexities [45], [46]. Also, in 1995, Admiral W.A. Owens was the first one to introduce the concept of SoS and highlight the importance of its development in the military [47].

In 1996, Manthorpe, Jr. proposed to link command, control, computers, communication, and information (C4I) with intelligence, surveillance, and reconnaissance (ISR) to join the SoS in order to achieve "dominant battlespace awareness" [6].

Also, in 1996, Maier, who is still considered to be one of the most influential contributors to the study of the SoS field, proposed for the first time to use the characterization approach to distinguish "monolithic" systems from SoS. These characteristics include "operational independence of the elements, managerial independence of the elements, evolutionary development, emergent behavior, and geographical distribution" [8]. In 1997, Kotov provided one of the most precise definitions of the SoS in the application of information technology. Also, he was one of the first scientists to attempt to model and synthesize SoS [5].

In 1998, Maier published an updated version of his 1996 "Architecting Principles of Systems-of-Systems," where he offered a new definition of SoS: "A system-of-systems is an assemblage of components which individually may be regarded as systems, and which possesses two additional properties: Operational Independence of the Components (and) Managerial Independence of the Components." [14].

Also, in 1998, Luskasik attempted to apply SoS approach in the educational context [7]. In 2000, Pei introduced a new concept of "system-of-systems integration" (SOSI) which gave the ability "to pursue development, integration, interoperability, and optimization of systems" to reach better results in "future battlefield scenarios" [48].

In 2001, four major contributors published their respective works to address SoS development. They included Sage and Cupan; Cook; Carlock and Fenton; and Shenhar. Sage and Cupan proposed to use principles of "new federalism" to provide a framework for the SoSE [4]. Cook examined SoS and described a distinction between "monolithic" systems and SoS based on "system attributes and acquisition approaches" [49]. He showed that constituent systems of SoS are acquired through separate processes [49]. Carlock and Fenton suggested on joining "traditional systems engineering activities with enterprise activities of strategic planning and investment analysis" [50]. They called this type of engineering "enterprise Systems of Systems engineering" [50]. Finally, Shenhar continued to expand the concept of "array" from his previous work published in 1994 [51].

In 2003, Keating, et al., presented a significant comparative study of SE and SoSE and provided guidelines for several key phases such as design, deployment, operation, and transformation of SoS [3]. Also in 2003, Chen and Clothier [52] published work addressing the need for a SoSE framework. They suggested advancing SE practices beyond traditional project level to focus on "organizational context" [52].

Another major contribution came from Bar-Yam and his study group in 2004. He examined applications of SoS in different fields and suggested adding characteristics as opposed to definitions to provide a more comprehensive view of SoS [9]. In 2005, there were numerous papers published on the topic of SoS, but we believe that the most significant inputs were produced by Jamshidi; Lane and Valerdi; and DeLaurentis. Jamshidi applied a definitional approach to SoS by collecting different definitions from various fields [53]. Lane and Valerdi used a comparative approach to analyze SoS definitions and concepts in the "cost models" context [15]. DeLaurentis described various traits within the transportation domain of SoS, and suggested the need to continue the search for a "holistic framework" and methodology [10].

In 2006, Boardman and Sauser published a paper that outlined the characterization approach to SoS. They identified patterns and differences in over 40 SoS definitions. By comparing these patterns and differences against previously identified patterns of other systems, they translated them into a comprehensive overview of five distinguishing characteristics of SoS. The characteristics are: autonomy, belonging, connectivity, diversity, and emergence [11].

In 2008, the first two books dedicated to SoS were introduced by Jamshidi [54], [55]. These works covered a wide variety of SoS topics. In addition to individual research efforts, many professional organizations, universities, government agencies, and non-forprofit organizations contributed to the advancement of understanding in studies of SoSE.

In 2001, the Department of Defense (DoD) introduced the following version of Army Software Blocking Policy (ASBP) v11.4E. It created a mechanism responsible for ensuring that "system developments" would be harmonized with the "execution of the program." It defined SoS as "a collection of systems that share/exchange information which interact synergistically" [56].

Also, in 2001, another department within the DoD, published a report in which it used "Family of Systems" (FoS) in a military context to describe SoS as a "set of arrangement of interdependent systems that can be arranged or interconnected in various ways to provide different capabilities" [57].

In 2004, another standard came from the DoD as part of Defense Acquisition Guidebook, where one of the sections was dedicated to a provision of guidelines during the acquisition phase of SoS [58].

In 2005, four major events took place—System of Systems Engineering Center of Excellence (SoSECE) 1st Conference, System of Systems Report from Purdue University, the establishment of National Center for System of Systems Engineering (NCSoSE) at Old Dominion University, and the United States Air Force (USAF) Report on SoSE.

In 2006, SoS Navigator was published by Carnegie Mellon University, and the first IEEE Conference dedicated to SoSE (IEEE SoSE) was held in Los Angeles, CA. Also, the IEEE Systems Journal was formed to continue the education of the scientific community about significant developments in the study of SoSE.

In 2007, the DoD published their System of Systems Engineering Guide: Considerations for Systems Engineering in a System of Systems Environment [46]. The guide describes the characteristics of SoS environments and identifies complexities of SoS systems engineering (SoS SE).

Also, in 2007, DeLaurentis et al., proposed to create an International Consortium for System of Systems (ICSOS) to examine the problems and solution strategies related to SoS and SoSE [59].

## C. *Analysis of literature review*

Table 6.1 presents a summary of the literature review including relevant works of Keating, et al. [3], Maier [14], DeLaurentis [10], Bar-Yam, et al. [9],

*Table 6.1* Major Areas of SE and SoSE

| | SE | SoSE |
|---|---|---|
| **Focus** | Single<br>Complex<br>System | Multiple<br>Integrated<br>Complex Systems |
| **Objective** | Optimization | Satisficing, Sustainment |
| **Boundaries** | Static | Dynamic |
| **Problem** | Defined | Emergent |
| **Structure** | Hierarchical | Network |
| **Goals** | Unitary | Pluralistic |
| **Approach** | Process | Methodology |
| **Timeframe** | System Life Cycle | Continuous |
| **Centricity** | Platform | Network |
| **Tools** | Many | Few |
| **Management Framework** | Established | ? |

and the research group at SoSECE [60] on the major aspects of both SE and SoSE. Aside from major differences, there is a gap in the literature in regards to the management framework of SoSE.

## *System of Systems Engineering (SOSE) Management*

While the scope of engineering and managing systems has changed dramatically and become a significant challenge in our ability to achieve success [61], fundamental to understanding the context of any system is the necessity to distinguish between the system type and its strategic intent, as well as its systems engineering and managerial problems [62]. Therefore, no single approach can solve these emerging problems, and thus no one strategy is best for all projects [63]. As already discussed, there can be great differences among systems and among the processes of their creation. Accordingly, it is a common practice for most organizations to use some kind of a project or systems engineering management classification or categorization framework, either explicitly or implicitly [64]. For example, Ahituv and Neumann [65]; Blake [66]; Steele [67]; and Wheelwright and Clark [68] were some of the earliest to propose frameworks for the distinction among systems and projects. While several others have suggested additional frameworks in an attempt to categorize and distinguish between different project types (e.g., Bubshait and Selen [69]; Floricel and Miller [70]; Pich et al., [71]; Shenhar and Dvir [12]; Turner and Cochrane [72]; Youker [73]), much of this literature has been focused on a single industry and often on small projects (Soderlund [74]; Tatikonda and Rosenthal [75]). Thus, the application of these frameworks to SoS has limited external validity. Furthermore, while most organizations use classification or categorization systems [64], only few of these systems are presently grounded in academicoriented empirical research.

While the literature in SoS and SoSE is expanding rapidly, we still do not have an established body of knowledge. We also lack, as our literature reviews indicated, a management framework for guiding us in our understanding of these complex systems. We will present a SoSE management framework that brings together a leading approach to describing SoS (i.e. characterization) and a fundamental trait of SoS (i.e. networks). In the next sections we will describe our characterization and network management models individually. We will then bring them together to form a SoSE Management Framework.

## *A. SoS characteristics*

We have chosen to use the SoS characterization of Boardman and Sauser [11], [76] because these characteristics are based on a review of over 40 definitions of SoS from the literature. A summary of these characteristics is as follows.

Autonomy—The ability of a system as part of SoS to make independent choices. This includes managerial and operational independence while accomplishing the purpose of SoS:

> The reason a system exists is to be free to pursue its purpose. That freedom always comes with constraints, of course. But those constraints cannot be permitted to overwhelm or violate its nature to perform. Were this to be the case, the system of necessity would be abandoned and another found to take its place. True, any given system may fail to fulfill its purpose, but not for reasons of autonomy. More likely it is ineffectiveness, efficiency, or even unethical behavior. The same cannot be said of a part that is integral to a system. That part is chosen—designed or procured—for a given purpose, just as a system is, but it is deliberately chosen for the reason of serving the purpose of the whole system [11].

Belonging—Constituent systems have the right and ability to choose to belong to SoS. The choice is based on their own needs, beliefs, or fulfillment:

> Part of the persuasion comes from the argument that the achievement of the SoS purpose is exactly why the system was brought into being, but constraints at the time of its origination required a lesser target to be set. In other words, the new 'supra' purpose enfolds the system's original purpose. And what is more, the existence of the SoS will enhance the value of the system's purpose, exalt the role of

the system, whose belonging makes achievement of the supra purpose more likely and more effective. But that belonging does mean partness for the autonomous system. This autonomous legacy system now exhibits both partness and wholeness [11].

Connectivity—The ability to stay connected to other constituent systems:

Now we are faced with the need to create connectivity, or in other words achieve interoperability, amongst the legacy systems and possibly additions of new systems to SoS It calls for a dynamic determination of connectivity, with interfaces and links forming and vanishing as the need arises. Thus, the ability of constituent systems to remain autonomous proves essential, for only then can they hope to make real-time connections on behalf of the SoS to enable it achieve and sustain its capabilities [11].

Diversity—Evidence of visible heterogeneity:

A SoS should, out of necessity, be incredibly diverse in its capability as a system compared to the rather limited functionality of a constituent system, limited by design. It seems to us that there is a fundamental distinction to be made between requirements-driven design for a conventional system based on its defined scope, and a capabilities-based SoS that must exhibit a huge variety of functions, on an as-needed basis, in order to respond to rampant uncertainty, persistent surprise, and disruptive innovation [11].

Emergence—Formation of new properties as a result of developmental or evolutionary process:

In a system, emergence is deliberately and intentionally designed in. What's more, unintended consequences, i.e., unpleasant or painful emergent behavior is tested out, as far as possible. With an SoS, emergent behavior dare not be restricted to what can be foreseen or deliberately designed in, even if this risks greater unintended consequences, though of course these can still be tested for. A SoS must be rich in emergence because it may not be obvious what tactical functionality is required to achieve broad capability. Instead, a SoS has emergent capability designed into it by virtue of the other factors: preservation of constituent systems autonomy, choosing to belong, enriched connectivity, and commitment to diversity of SoS manifestations and behavior. The challenge for the SoS designer is to know, or learn how, as the SoS progresses through its series of stable states, to create a climate in which emergence can flourish, and an agility to quickly detect and destroy unintended behaviors [11].

According to Sauser and Boardman [77], these five distinguishing characteristics of SoS are influenced by opposing forces with different degrees of strength within each of them. Figure 6.3 depicts the various forces' that

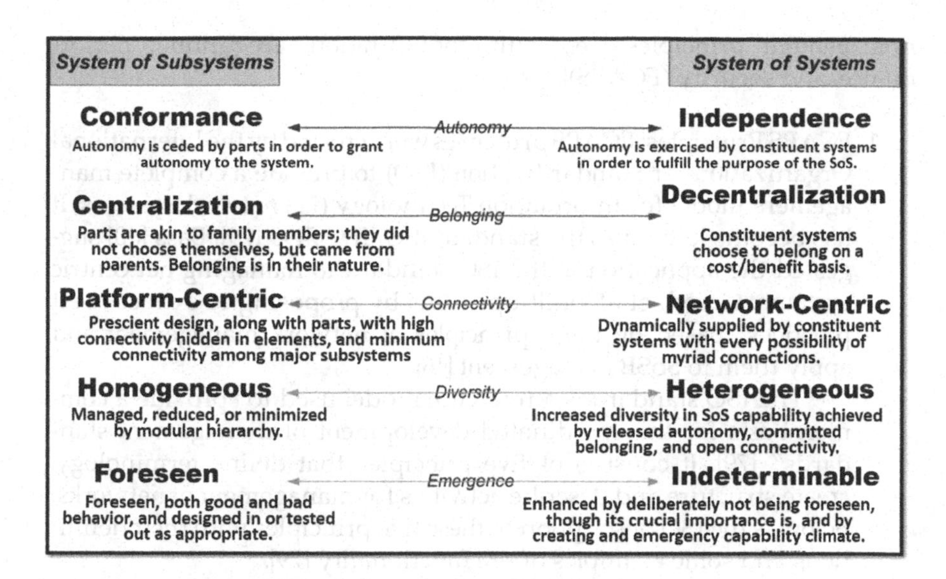


*Figure 6.3* Distinguishing characteristics of SoS and their opposing forces [78].

interact within these characteristics. We will later use these opposing forces to better define our framework.

Likewise, we contend that these characteristics are not independent but meaningfully interdependent, as they extend from one extremity to the other for each (i.e., paradox). To understand the interdependence of these characteristics is to begin to gain access to how they interlock. For example, together the level of autonomy may determine the degree of belonging, which will affect the extent of connectivity and possibly restrict the diversity (of elements) and maybe the emergent properties (of the system). However, equally a shift in diversity may have an effect in belonging and hence connectivity leading to a rising of the autonomy level and consequential effect on emergence. Our only present recourse is to nominate fluxes in the dynamics of these characteristics based on the experience of designed systems, and latterly SoS. Thus, we reserve this discovery for future research.

## B. *Network management*

In our review of the literature, we established that SoS can be viewed as a network. Shenhar [13]; Maier [14]; Lane and Valerdi [15]; and DeLaurentis [10] all provided examples of collaborative SoS such as the Internet, global communication networks, transportation networks, etc., which can be managed as networks. Therefore, we extracted the management practices of how networks are governed from already established network management principles [i.e., fault, configuration, accounting, performance, and security (FCAPS)].

1. FCAPS Principles: FCAPS principles were created by the International Organization for Standardization (ISO) to provide a complete management model for Information Technology (IT) network systems. It has become known as the standard ISO/IEC 7498 [79]. Bass [80] suggested the application of the ISO standard to managing net-centric systems. Gorod, et al. built upon this by proposing to extract "best practices" based on these principles of network management and apply them to SoSE management [16].

   The ISO standard is a reference model used to "provide a common basis for the coordinated development of management standards" [79]. It consists of five principles that define terminology, create structure and describe activities for management of networks. Below is the ISO description of these five principles with their definitions and some examples of the functionality [79].

   Fault Management (FM)—Encompasses fault detection, isolation and the correction of abnormal operation of the Open Systems Interconnection Environment (OSIE). Functionality of FM:

   1. Maintains and examines error logs;
   2. accepts and acts upon error detection notifications;
   3. traces and identifies faults;
   4. carries out sequences of diagnostic tests;
   5. corrects faults.

   Configuration Management (CM)—Identifies, exercises control over, collects data from, and provides data to open systems for the purpose of preparing for, initializing, starting, providing for the continuous operation of, and terminating interconnection services. Functionality of CM includes:

   1. setting the parameters that control the routine operation of the open system;
   2. associating names with managed objects and sets of managed objects;
   3. initializing and closing down managed objects;
   4. collecting information on demand about the current condition of the open system;
   5. obtaining announcements of significant changes in the condition of the open system;
   6. changing the configuration of the open system.

   Accounting Management (AM)—Enables charges to be established for the use of resources in the OSIE, and for costs to be identified for the use of those resources. Functionality of AM includes:

1. informing users of costs or resources consumed;
2. enabling accounting limits to be set and tariff schedules to be associated with the use of resources;
3. enabling costs to be combined where multiple resources are invoked to achieve a given communication objective.

Performance Management (PM)—Enables the behavior of resources in the OSIE and the effectiveness of communication activities to be evaluated. Functionality of PM includes:

1. gathering statistical data;
2. maintaining and examining logs of system state histories;
3. determining system performance under natural and artificial conditions;
4. altering system modes of operation for the purpose of

conducting performance management activities. Security Management (SM)—Support the application of security policies. Functionality of SM includes:

1. creation, deletion, and control of security services and mechanisms;
2. distribution of security relevant information;
3. reporting of security relevant events.

2. Conceptual Areas of SoSE Management: While the description and definitions above apply to Information Technology (IT) networks, we have theorized that certain values of the FCAPS principles can be abstracted in order to apply to the SoSE domain as depicted in Figure 6.4 [16]. The resulting conceptual areas serve as a foundation for developing a framework for SoSE management.

   These SoSE Management Conceptual Areas can be described in the following manner.

   Risk Management—Monitor, identify, assess, analyze, and mitigate risk encountered in the SoS.

   Configuration Management—Direct and coordinate through functioning and software management. Performance Management—Monitor and measure performance of SoS for it to be maintained at an appropriate level. Policy Management—Provide SoS access to authorized processes and protect SoS from illegal access.

   Business Management—Coordinate and allocate SoS assets based on use and utilization information of systems in the SoS.

3. SoSE Management Matrix (SoSEMM): These five foundations of SoSE Management play an important role in the process of creating a SoSEMM. This should provide guidance in the process of increasing an overall effectiveness of SoSE management practices and the creation of a much needed management framework. The relationship between distinguishing characteristics of SoSE and

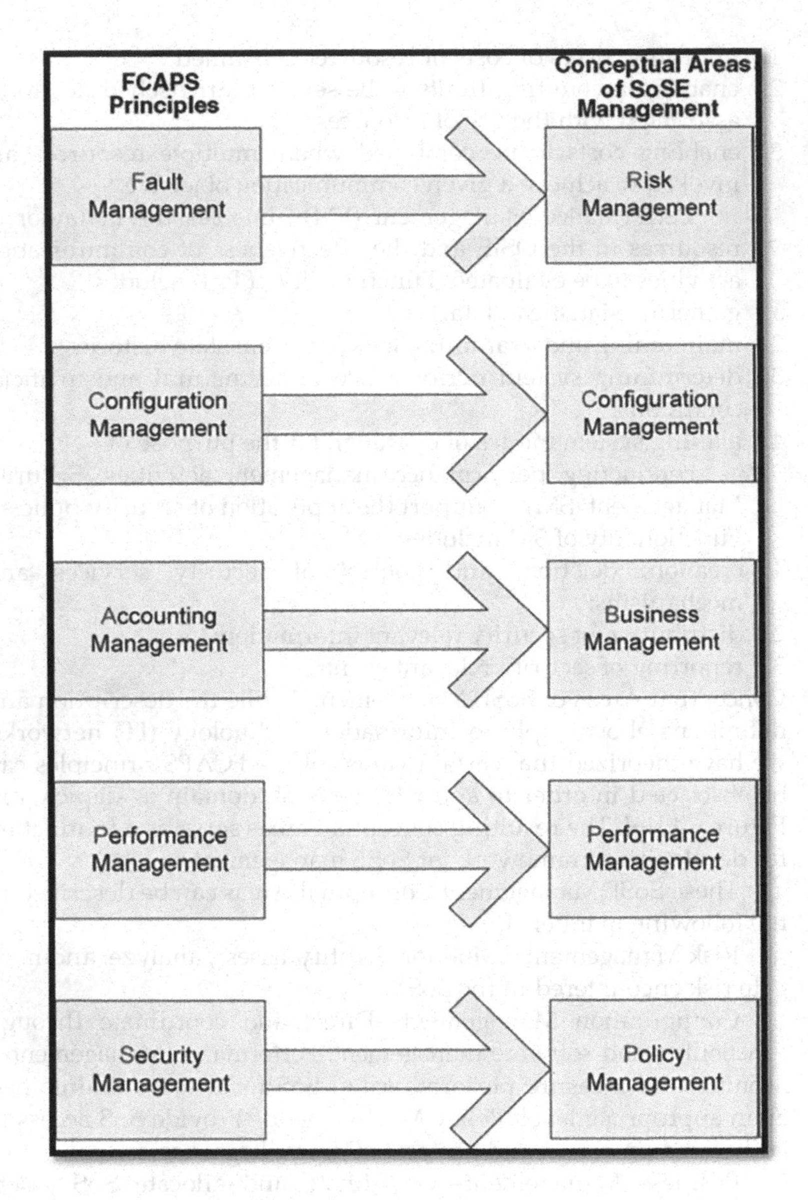


*Figure 6.4* Process of abstracting FCAPS to SoSE domain, adopted from [16].

management processes is represented by the SoSEMM which is depicted in Figure 6.5.

4. SoSE Management Framework: We propose utilizing the SoSEMM for the purpose of creating an effective SoSE management framework. By using the five distinguishing characteristics described in Section III-A and modified FCAPS principles (SoSE conceptual

| SoSEMM | Risk Management | Configuration Management | Performance Management | Policy Management | Business Management |
|---|---|---|---|---|---|
| Autonomy | Preserves autonomous capabilities | Governs autonomous behavior | "Satisficing" constituent system performance in order to meet SoS goals | Ensures security of constituent systems | Provides information on autonomous resources |
| Belonging | Provides environmental perspective | Unifies configuration of constituent systems | Allows cost/benefit analysis | Restricts/allows constituent systems access to the SoS | Provides information on constituent system functions |
| Connectivity | Maintains acceptable integration | Establishes and maintains consistency in integration | "Satisficing" to maximum capability | Protects against unauthorized integration | Tracks connected resources |
| Diversity | Evaluates diverse activities | Governs diverse behavior | "Satisficing" diverse activities | Restricts/allows diverse behavior | Provides information on inventory of diverse resources |
| Emergence | Verifies and validates achievement of new behaviors | Shapes and bounds emergent behavior | Permits "Satisficing" path of emergence | Influences emergent behavior | Tracks emergent capabilities |

***Figure 6.5*** SoSEMM adopted from [16].

areas) outlined in Section III-B2, the framework is created that possesses four essential functions. First, it provides us with the ability to describe the current overall context of a SoS. Second, it allows us to identify dynamic processes within the five distinguishing characteristics (individual gauges) while various forces interact within a SoS. Third, it facilitates feedback processes about the first and second functions. This third function directs us to the fourth one that encompasses the conceptual areas (triggers) and enables us to govern SoS. Figure 6.6 depicts these four functions followed by Figure 6.7 which reflects proposed SoSE management framework that includes the SoSEMM. The SoSEMM provides us with the ability to identify different relationships between distinguishing characteristics and Conceptual Areas. It serves as the necessary link to recognize and govern various processes within SoS. Section IV reflects the case study of the Integrated Deepwater System of Systems (IDS) as an example of the application of the proposed SoSE management framework.

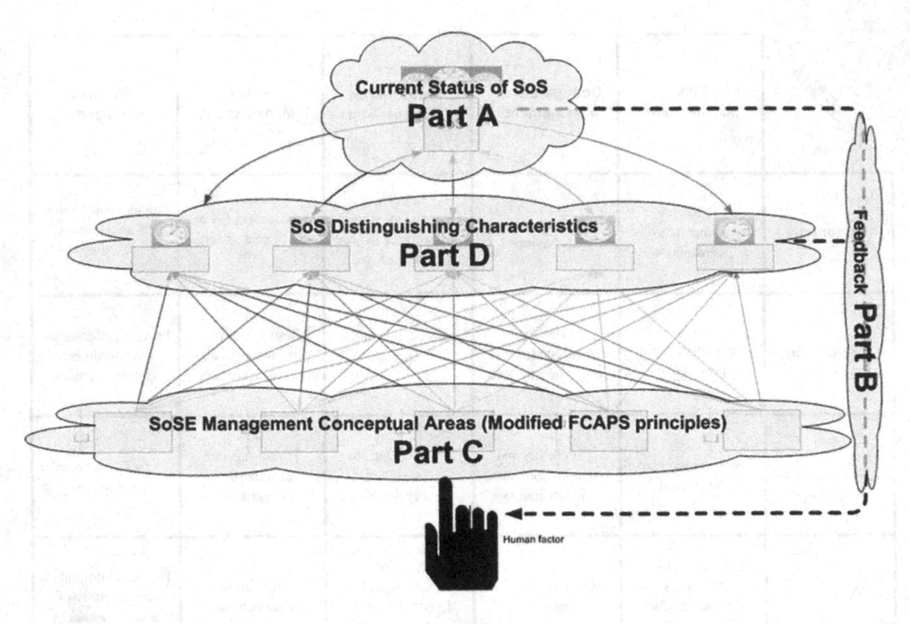


*Figure 6.6* Four functions of SoSE management framework.

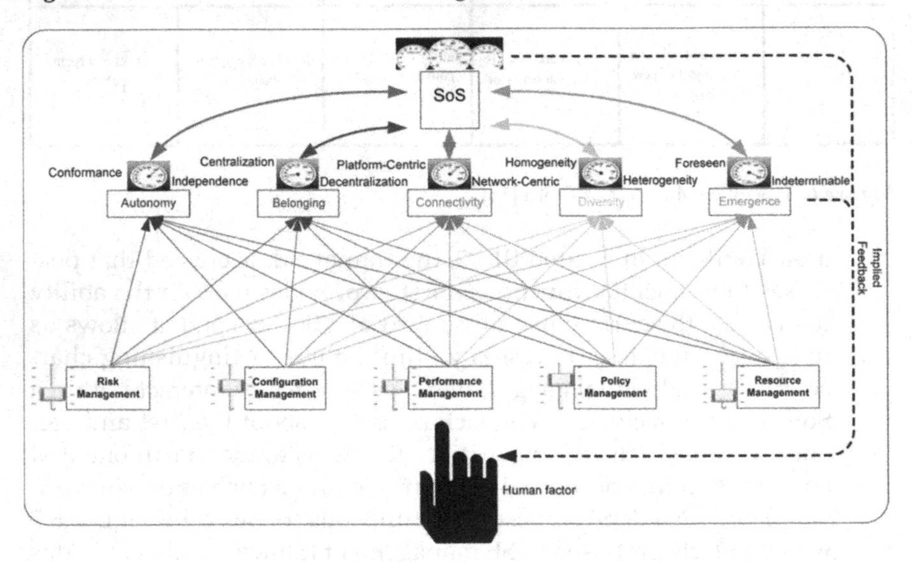


*Figure 6.7* SoSE management framework.

Figure 6.6 depicts four functions in the following order.

- Part A—The first function indicates the current overall status of SoS.
- Part B—The second function facilitates the flow of information (feedback process) about Part A, Part D to Part C.

- Part C—The third function receives valuable input from Part B and develops necessary policies to influence the five distinguishing characteristics in Part D.
- Part D—The fourth function illustrates the interaction between various forces within each of the five distinguishing characteristics of SoS.

In summary, when we have obtained information about the current overall status through the feedback process, it then becomes possible to utilize modified FCAPS principles to effectively govern SoS through the five distinguished characteristics. Figure 6.7 shows the final result of interaction of different processes within SoS and the functionality of the effective SoSE management framework. SoSEMM is located between Part C (Conceptual Areas) and Part D (Distinguishing Characteristics) and illustrates their relationship.

Based on a similar methodology, Dobson et al., [81] proposed to use "Autonomic control loop" method to effectively cope with the ever changing environment in the network communication context. "The ultimate vision of autonomic communication research is that of a networked world in which networks and associated devices and services will be able to work in a totally unsupervised manner, able to self-configure, self-monitor, self-adapt, and self-heal-the so called "self-properties." On the one hand, this will deliver networks capable of adapting their behaviors dynamically to meet the changing specific needs of individual users; on the other, it will dramatically decrease the complexity and associated costs currently involved in the effective and reliable deployment of networks and communication services" [81]. Figure 6.8 illustrates the "Autonomic control loop."

## *Case study of the Integrated Deepwater System (IDS)*

This case study was originally introduced in the authors' previous work [82] as an example of SoS. However, in this paper, we are demonstrating an application of the proposed framework in the SoSE domain using the IDS case study.

The Integrated Deepwater System (IDS) serves as an illustration of an SoS example [46]. IDS was initiated by the Coast Guard to modernize some of the outdated "aviation and surface platforms and shore facilities" in the early 1990's [83]. The Coast Guard directed its efforts to develop an "integrated approach to upgrade existing assets while transitioning to newer, more capable platforms with improved systems for command, control, communications, computers, intelligence, surveillance, and reconnaissance (C4ISR) and integrated logistics" [83]. The project has an

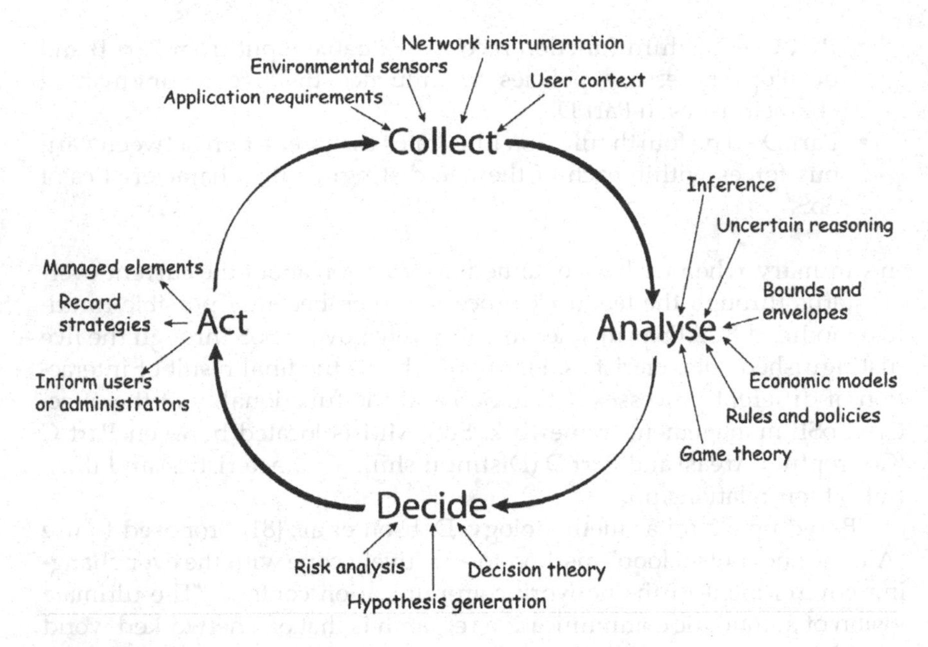


*Figure 6.8* Autonomic control loop [81].

objective to provide the Coast Guard "with a significantly improved ability to intercept, engage and deter those activities that pose a direct challenge to U.S. security" [83]. Having a $24 billion budget, IDS is scheduled to be completed in a period of 25 years and should include 91 cutters, 195 units of aircraft, C4ISR equipment and integrated logistics capabilities [83], [84]. Figure 6.9 provides an overview of operational plan for IDS.

*Figure 6.9* Integrated deep-water system © Can Stock Photo Inc./ yykkaa.

## A. Part A: current IDS status

There have been several major accomplishments in the IDS program. Recently, there have been significant results in the following four areas: sea, aviation, shore facilities, and systems engineering [85]. In aviation, 65 units of HH-65C re-engine helicopters and three units of HC-144A patrol aircrafts were completed. In maritime navigation, the first of the three National Security Cutters (NSC) is complete, the second is under construction and the third is approved to receive requisite supply materials. In shore facilities, there has been an upgrade process for two C4ISR command stations. In systems engineering, there has been a reduction in costs of $568 million by eliminating 82 wasteful programs. In addition, there has been a delivery of 88 C4ISR contracts of (DD250s) from the Maritime Domain Awareness Center to date [85].

Despite these accomplishments, IDS continues to face significant challenges. It has been severely criticized by many government agencies [84] for inadequate management and execution policies. According to the Department of Homeland Security Inspector General (DHS IG), the Government Accountability Office (GAO), the Defense Acquisition University (DAU), several Members of Congress from committees and subcommittees that oversee the Coast Guard, the initial time frame of 20 years has been shifted to 25 years and the cost of the overall project has increased from $17 billion to $24 billion [84].

## B. Part B: feedback

After we have obtained valuable information about the current overall status of IDS from different sources such as government agencies and other observers, there is a beginning of analytical processes to develop the necessary policies in influencing certain characteristics of a system for the purpose of effective governance of SoS. For example, there is a standard report prepared for Congress to provide relevant details about the current status of IDS with emphasis on accomplishments and setbacks in management and execution of the program [84]. Even though this report provides a comprehensive account of the facts, it also has the disadvantage of being time consuming. It does not address real time problems and should be supplemented by different sensor mechanisms that would immediately alarm the system about changing conditions.

## C. Part C: conceptual areas

As previously mentioned, once the information has been collected, it is possible to make an assessment of the situation based on the modified FCAPS principles. These modified principles will be discussed in detail below and include risk management, configuration management, performance management, policy management, and business management.

Risk Management—Risk management for the IDS is an assessment process of SoS objectives and different types of risks that might potentially jeopardize the achievement of these objectives. For example, it is absolutely necessary to collect and analyze information about unwanted emergent behavior, infrastructure, integration, and cost risks [46]. The IDS is already running the risk of being the over budget by $7 billion as Deepwater program's estimated total acquisition cost has increased from $17 billion to $24 billion [84].

Configuration Management—Configuration management for the IDS includes a process of identifying the current and likely mix of constituent systems within SoS that is needed for the user to accomplish its capability goals. It becomes necessary to properly establish flexible provisions that will encompass not only the evolution of various capabilities but also the addition of similar capabilities as they emerge in the entire defense structure. The Configuration Management structure has to find a meaningful method to correctly fit stakeholders and program office authorities and responsibilities [46].

Performance Management—Performance Management of the IDS is based on the comparison processes of technical requirements of various SoS constituents and their actual implementation. These measurements include assessment of the "external interfaces of the constituent systems, their compliance with the connectivity and communications protocols and standards, and their compatibility with the other constituent systems" [46]. In addition, it is necessary to view various maturity measurement mechanisms [i.e., Technology Maturity Levels (TML) and Systems Maturity Levels (SML)] in the configuration of the SoS context in response to the set of defined capabilities [46]. Policy Management—Policy Management of any SoS is developed in response to some form of management or execution defects. The acquisition policy of the IDS addresses the problem of ineffective human capital management. By consolidating various SoS constituents under management of one organization, such as the Coastal Guard, it now becomes possible to implement standard procedures for hiring and training purposes within IDS constituents to ensure most effective communication and security protocols [46].

Business Management—Business Management utilizes a holistic view on SoS and includes examination of a specific system from different angles in the context of overall SoS. In the IDS case study, it is important to identify various objectives posted for each individual constituent system. For example, vertical take-off-and-landing unmanned aerial vehicle (UAV) plays an important role within the aviation system because it fulfills specific tasks of reconnaissance from space. At the same time, it is limited in its capacity to carry massive cargo. The aviation system is only part of the overall IDS. Therefore, depending on the particular context, it becomes necessary to properly develop business management strategy to achieve mission objectives [83].

## *D. Part D: distinguishing characteristics and opposing forces*

The examination of the opposing forces/paradoxes within each characteristic will heuristically indicate a point where those forces meet.

**Autonomy**

- Conformance: Originally, the Coast Guard was the agency in charge of developing and implementing standard procedures for different constituent parts of the IDS program. Fairly recently, the committees and subcommittees of Congress that were responsible for the oversight of the Coast Guard efforts found numerous significant shortcomings in management and execution of the program. Since then, they proposed to develop a different strategy directed towards a more independent approach [84].
- Independence: It now becomes essential for the Coast Guard to give priority to an acquisition strategy that develops the individual acquisition projects within air, surface, or C4ISR and puts less focus on the grouping of projects in accordance to their respective program [83].

  Therefore, on the scale of paradoxes, the previously described chain of events reflects a slight shift to the right towards the independence side within the autonomy characteristic.

**Belonging**

- Centralization: The U.S. Coast Guard decided to construct IDS acquisition as a centralized system "under which a combination of new and modernized cutters, patrol boats, aircraft, along with associated C4ISR systems and logistics support, would be procured as a single, integrated package" [84].
- Decentralization: While IDS program has been "governed using a tightly focused system-like governance process, it is dominated by a diverse series of stakeholders associated with the various systems integrated into the SoS" [46].

  The IDS' belonging is characterized by loosely coupled, decentralized constituent systems sharing governing rules through conforming centralized guidance. As a result, the meeting point of the opposing forces within the belonging characteristic is closer to the decentralization side.

**Connectivity**

- Platform-Centric: According to the recent U.S. Department of Homeland Security report, "Coast Guard faces several challenges in implementing effectively its Deepwater C4ISR systems. Due to the limited oversight as well as unclear contract requirements, the agency cannot ensure that the contractor is making the best decisions toward accomplishing Deepwater IT goals. Insufficient C4ISR funding has restricted accomplishing the 'system-of-systems' objectives that are considered fundamental to Deepwater asset interoperability. Inadequate training and guidance hinder users from realizing the full potential of the C4ISR upgrades" [86]. However, Coast Guard is trying to address these issues [83].
- Network-Centric: By integrating C41SR capability the IDS attempts to "upgrade aging and increasingly obsolete legacy assets while transitioning to newer and more capable platforms—leading to an integrated and interoperable network-centric system of systems" [87]. Effective utilization of C4ISR capabilities will provide different federal, state, local and other government agencies with the ability to access information and tracks as needed in a secure and efficient manner.

  Therefore, the opposing forces of the connectivity in IDS case intersect closer to the network-centric side.

**Diversity**

- Homogeneous: Along with the acquisition of many new assets and systems, the IDS program "encompass other work, including, originally, the conversion of the Coast Guard's existing 49 Island-class 110-foot patrol boats into modernized, 123-foot patrol boats, so that these boats could remain in service until the delivery of replacement" [84].
- Heterogeneous: As mentioned earlier, the IDS program includes the acquisition of many new highly diversified assets and systems. They include different types of patrol boats, classes of cutters, small boats, various aircrafts, logistics tools, and C41SR systems. In addition, "fixed wing, rotary wing, and unmanned aircraft, and communications" [46] also contribute to the heterogeneity of IDS.

  The newly acquired assets and systems of IDS still have to coexist with the legacy systems that were not originally designed or developed to operate as part of SoS. Consequently, the point where homogeneous and heterogeneous forces meet is located fairly close to the heterogeneous side.

**Emergence**

- Foreseen: The Deepwater program is a "performance-based acquisition, meaning that it is structured around results to be achieved rather than the manner in which the work is performed. If performance-based acquisitions are not appropriately planned and structured, there is an increased risk that the government may receive products or services that are over cost estimates, delivered late, and of unacceptable quality" [84].

- Indeterminable: The objective of IDS is to transform the Coast Guard so it is emergent enough to cope with the constantly changing operating environment. According to Admiral T. W. Allen: "The Deepwater Program will provide more capable, interoperable assets that will enable our forces to close today's operational gaps and to perform their demanding missions more effectively, efficiently, and safely" [83].

As it currently stands, IDS leans slightly toward the indeterminability side within the emergence characteristic.

Based on the analysis above of the five distinguishing characteristics and their paradoxes, Figure 6.10 demonstrates a qualitative depiction of the meeting point of the various forces within IDS. While it is improbable to indicate the precise meeting point on this scale using the qualitative analysis just described, the authors are currently investigating quantitative approaches to making this determination.

Figure 6.11 depicts the relationship between governing processes and characteristics of IDS. The SoSEMM shows us ways to implement corrective action to effectively influence each of the five distinguishing characteristics.

The IDS case study demonstrates an application of the proposed framework in the SoSE domain. Our goal includes engineering systems which can be simultaneously emergent enough to keep up and cope with the changing operating environment, yet are also effectively governable. The framework guides us in achieving this goal.

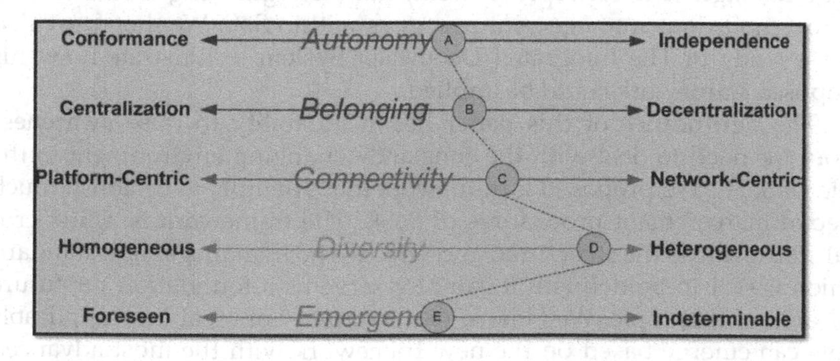


*Figure 6.10* Five distinguishing characteristics of the IDS.

| SoSEMM | Risk Management | Configuration Management | Performance Management | Policy Management | Business Management |
|---|---|---|---|---|---|
| **Autonomy** | Permits preservation of autonomous capabilities of IDS's constituent systems | Provides guidelines to effectively govern autonomous behavior of the IDS' constituent systems | Monitors IDS' constituent systems performance in order to meet SoS goals | Helps establish the extent of the autonomy for the constituent systems of IDS | Provides information which enables influencing the autonomy of the IDS' constituent systems |
| **Belonging** | Permits effective cooperation among IDS constituent systems toward a common goal | Provides direction and coordination to unify configuration of the IDS' constituent systems | Monitors IDS' constituent systems performance in order to evaluate their contributions | Helps to decipher the level of belonging of the constituent systems of the IDS | Provides information showing the extent to which constituent systems belong to IDS |
| **Connectivity** | Permits to maintain acceptable integration of the IDS's constituent systems | Establishes and maintains consistency in integration of IDS' assets (e.g., protocols, standards, etc. | Monitors the level of interoperability of IDS' assets | Helps provide identification and authentication of the constituent systems to prevent unauthorized integration | Tracks connected resources of IDS. Capabilities provide an ability to access information |
| **Diversity** | Helps to evaluate diverse behavior of the IDS's constituent systems | Helps to govern diverse nature of IDS' constituent systems | Monitors the extent of diversity of IDS's assets related to performance | Restricts/permits diverse behavior of IDS' constituent systems | Provides information on IDS' highly diversified assets and systems |
| **Emergence** | Substantiates achievement of new behaviors based on monitoring, identifying, assessing, analyzing, and mitigating risk | Shapes and bounds emergent behavior in order for IDS to stay within the scope of its goals (e.g., protocols, standards, doctrines, etc.) | Monitors the extent of emergence of IDS | Bounds emergent behavior in order for IDS to operate effectively, efficiently, and safely | Tracks new emergent properties both intended and unintended |

*Figure 6.11* Application of the SoSEMM in the IDS.

## Conclusion

The review of SoS literature illustrates the necessity to create an SoSE management framework based on the demands of constant technological progress in the complex dynamic environment. After careful analysis of the history and evolution of SoS, several authors identified SoS as a network and suggested extracting "best practices" of network management to SoSE domain.

To accomplish the objective of creating an effective SoSE management framework, we proposed to utilize the modified FCAPS principles (SoSE Management Conceptual Areas) and distinguishing characteristics to govern various management processes within SoS. We then provided a case study of The Integrated Deepwater System to illustrate how this proposed framework could be applied.

The significance of this paper lies in its ability to raise awareness about the need to deal with the constantly changing environment in the field of SoSE. We proposed a framework that attempts to establish much needed management procedures of SoSE. The framework is a first crucial step towards the structured and effective SoSE management standard which is still to be defined. It can also serve as a foundation for future research of this topic. We believe that a number of "real life" applicable tools can emerge based on the new framework with the most advanced ones becoming part of a "SoSE tool-set."

## *References*

1. C. Francois, "Systemics and cybernetics in a historical perspective,"Syst. Res. Behavioral Sci., vol. 16, no. 3, pp. 203–219, 1999.
2. IEEE Standard for Application and Management of the Systems Engineering Process, IEEE 1220, 2005.
3. C. Keating, R. Rogers, R. Unal, D. Dryer, A. Sousa-Poza, R. Safford, W. Peterson, and G. Rabaldi, "Systems of systems engineering," Eng. Management J., vol. 15, no. 3, pp. 36–45, 2003.
4. A. P. Sage and C. D. Cuppan, "On the systems engineering and management of systems of systems and federations of systems," Inf., Knowl., Syst. Management, vol. 2, pp. 325–345, 2001.
5. V. Kotov, "Systems of systems as communicating structures," Hewlett Packard, HPL-97-124, 1997.
6. W. H. Manthorpe, "The emerging joint system of system: A systems engineering challenge and opportunity for APL," in John Hopkins APL Techn. Dig., 1996, vol. 17, no. 3, pp. 305–310.
7. S. J. Luskasik, "System, systems of systems, and the education of engineers," Artif. Intell. Eng. Des., Analy., Manuf., vol. 12, no. 1, pp. 55–60, 1998.
8. M. Maier, "Architecting principles of systems-of-systems," presented at the 6th Ann. Int. Symp. Int. Council Syst. Eng., Boston, MA, 1996.
9. Y. Bar-Yam, M. A. Allison, R. Batdorf, H. Chen, H. Generazio, H. Singh, and S. Tucker, "The characteristics and emerging behaviors system of systems," NECSI: Complex Physical, Biological and Social Systems Project, 2004.
10. D. DeLaurentis, "Understanding transportation as a system of systems design problem," presented at the 43rd AIAA Aerosp. Sci. Meet., Reno, NV, 2005.
11. J. Boardman and B. Sauser, "System of systems—The meaning of OF," presented at the IEEE Int. Syst. Syst. Conf., Los Angeles, CA, 2006.
12. A. Shenhar and D. Dvir, Reinventing Project Management: The Dia-mond Approach to Successful Growth and Innovation. Boston, MA: Harvard Business School, 2007.
13. A. Shenhar, "A new systems engineering taxonomy," in Proc. 4th Int. Symp. Nat. Council Syst. Eng., 1994, pp. 261–276.
14. M. Maier, "Architecting principles for system-of-systems," Syst. Eng., vol. 1, no. 4, pp. 267–284, 1998.
15. J. A. Lane and R. Valerdi, "Synthesizing SoS concepts for use in cost estimation," presented at the IEEE Conf. Syst., Man, Cybern., Waikoloa, HI, 2005.
16. A. Gorod, R. Gove, B. Sauser, and J. Boardman, "System of systems management: A network management approach," presented at the IEEE Int. Conf. Syst. Syst. Eng., San Antonio, TX, 2007.
17. B. Sauser, "Toward mission assurance: A framework for systems engineering management," Syst. Eng., vol. 9, no. 3, pp. 213–227, 2006.
18. J. H. Brill, "Systems engineering—A retrospective view," Syst. Eng., vol. 1, pp. 258–266, 1998.
19. A. D. Hall, A Methodology for Systems Engineering, V. Nostand, Ed. Princeton, NJ, 1962.
20. H. Goode and R. Machol, Systems Engineering. New York: McGraw-Hill, 1957.

21. E. W. Engstrom, "Systems engineering—A growing concept," Elect. Eng., vol. 76, pp. 113–116, 1957.
22. A. D. Hall, Metasystems Methodology. New York: Pergamon, 1989.
23. S. Shinners, Techniques of Systems Engineering. New York: McGraw-Hill, 1967.
24. H. Chestnut, Systems Engineering Tools. New York: Wiley, 1965.
25. R. F. Miles, Systems Concepts. New York: Wiley, 1973.
26. W. P. Chase, Management of Systems Engineering. Malabar, FL: Robert Krieger, 1974.
27. W. Wymore, Systems Engineering Methodology for Interdisciplinary Teams. New York: Wiley, 1976.
28. B. Blanchard and W. Fabrycky, Systems Engineering and Analysis. Englewood Cliffs, NJ: Prentice-Hall, 1981.
29. A. P. Sage, Systems Management for Information Technology and Software Engineering. New York: Wiley, 1995.
30. AFSC, Washington, DC, "Manual 375-5, systems engineering management procedures," 1966.
31. USAF, "MIL-STD-499 systems engineering management," 1969.
32. USAF, "MIL-STD-499A engineering management," 1974.
33. DoD, "Filed Manual 770–78 systems engineering," 1979.
34. DSMC, "Systems engineering management guide," 1983.
35. USAF, "MIL-STD-499B systems engineering," 1992.
36. EIA, "Interim Standard 632 Systems Engineering," 1994.
37. K. E. Boulding, "General systems theory—The skeleton of science," Manag. Sci., vol. 2, no. 3, p. 197, 1956.
38. M. C. Jackson and P. Keys, "Towards a system of systems methodologies," J. Oper. Res. Soc., vol. 35, no. 6, pp. 473–486, 1984.
39. R. Ackoff, "Towards a system of systems concepts," Manag. Sci., vol. 17, no. 11, pp. 661–672, 1971.
40. F. Jacob, The Logic of Living Systems. London, U.K.: Allen Lane, 1974.
41. G.P.O, U.S., "Restructuring of the strategic defense initiative (SDI) program," 1989.
42. H. Eisner, J. Marciniak, and R. McMillan, "Computer-aided system of systems (C2) engineering," presented at the IEEE Int. Conf. Syst., Man Cybern., Charlottesville, VA, 1991.
43. H. Eisner, "RCASSE: Rapid computer-aided systems of systems (S2) engineering," in 3rd Int. Symp. Nat. Council Syst. Eng., 1993, pp. 267–273.
44. A. Shenhar and B. Sauser, "Systems engineering management: The multidisciplinary discipline," in Handbook of Systems Engineering and Management, 2nd ed. New York: Wiley, 2008.
45. J. H. Holland, Hidden Order: How Adaptation Builds Complexity. Reading, MA: Addison-Wesley, 1995.
46. DoD, "System of systems, systems engineering guide: Considerations for systems engineering in system of systems environment," 2007.
47. A. W. A. Owens, "The emerging U.S. system of systems," in Dominant Battlespace Knowledge, S. Johnson and M. Libicki, Eds. Washington, DC: NDU Press, 1995.
48. R. Pei, "System of systems integration (SoSI)—A smart way of acquiring army C412WS systems," in Summer Comput. Simulation Conf., 2000, pp. 574–579.

49. S. C. Cook, "On the acquisition of systems of systems," presented at the INCOSE Ann. Symp., Melbourne, Australia, 2001.
50. P. G. Carlock and R. E. Fenton, "System of systems (SoS) enterprise systems engineering for information-intensive organizations," Syst. Eng., vol. 4, no. 4, pp. 242–261, 2001.
51. A. Shenhar, "One size does not fit all projects: Exploring classical contingency domains," Manag. Sci., vol. 47, no. 3, pp. 394–414, 2001.
52. P. Chen and J. Clothier, "Advancing systems engineering for systems-of-systems challenges," Syst. Eng., vol. 6, no. 3, pp. 170–181, 2003.
53. M. Jamshidi, System of systems engineering—A definition. Piscataway, NJ: IEEE SMC, 2005.
54. M. Jamshidi, Ed., System of System Engineering—Innovations for the 21st Century. Hoboken, NJ: Wiley, 2008.
55. M. Jamshidi, Ed., System of Systems - Principles and Applications. Boca Raton, FL: Taylor & Francis Group, 2008.
56. DoD, "Army software blocking policy: v.11.4E," 2001.
57. JCIDS, "Joint capabilities integration and development instructions," 2001.
58. DAU, "Defense acquisition guidebook," 2004.
59. D. DeLaurentis, C. Dickerson, M. DiMario, P. Gartz, M. M. Jamshidi, S. Nahavandi, A. P. Sage, E. B. Sloane, and D. R. Walker, "A case for an international consortium on system-of-systems engineering," IEEE Syst. J., vol. 1, no. 1, pp. 68–73, Sep. 2007.
60. SoSECE, "System of systems engineering center of excellence," 2006 [Online]. Available: http://www.sosece.org/
61. C. N. Calvano and P. John, "Systems engineering in an age of com-plexity," Syst. Eng., vol. 7, no. 1, pp. 25–34, 2004.
62. A. Shenhar and B. Sauser, "Systems engineering management: The multidisciplinary discipline," in Handbook of Systems Engineering and Management, 2nd ed. New York: Wiley, 2008.
63. A. Shenhar, "One size does not fit all projects: Exploring classical contingency domains," Manag. Sci., vol. 47, no. 3, pp. 394–414, 2001.
64. L. Crawford, J. B. Hobbs, and J. R. Turner, Project Categorization Systems. Newtown Square, PA: Project Management Institute, 2004.
65. N. Ahituv and S. Neumann, "A flexible approach to information system development," MIS Q, pp. 69–78, Jun. 1984.
66. S. B. Blake, Managing for Responsive Research and Development. San Francisco, CA: Freeman, 1978.
67. L. W. Steele, Innovation in Big Business. New York: Elsevier Publishing Company, 1975.
68. S. C. Wheelwright and K. B. Clark, Revolutionizing Product Development. New York: The Free Press, 1992.
69. K. A. Bubshait and W. J. Selen, "Project characteristics that influence the implementation of project management techniques: A survey," Project Manag. J., vol. 23, no. 2, pp. 43–47, 1992.
70. S. Floricel and R. Miller, "Strategizing for anticipating risks and turbulance in large-scale engineering projects," Int. J. Project Manag., vol. 19, pp. 445–455, 2001.
71. M. T. Pich, C. H. Loch, and A. D. Meyer, "On uncertainty, ambiguity, and complexity in project management," Manag. Sci., vol. 48, no. 8, pp. 1008–1023, 2002.

72. J. R. Turner and R. A. Cochrane, "Goals-and-methods matrix: Coping with projects with ill defined goals and/or methods of achieving them," Int. J. Project Manag., vol. 11, no. 2, pp. 93–101, 1993.
73. R. Youker, "The difference between different types of projects (revised)," presented at the PMI 30th Ann. Seminar Symp., Philadelphia, PA, 2002.
74. J. Soderlund, "Developing project competence: Empirical regularities in competitive project operations," Int. J. Project Manag., vol. 9, no. 4, pp. 451–480, 2005.
75. M. V. Tatikonda and S. R. Rosenthal, "Technology novelty, project complexity, and product development project execution success: A deeper look at task uncertainty in product innovation," IEEE Trans. Eng. Manag., vol. 47, no. 1, pp. 74–87, 2000.
76. J. Boardman and B. Sauser, Systems Thinking: Coping With 21st Century Problems. Boca Raton, FL: Taylor & Francis Group, 2008.
77. B. Sauser and J. Boardman, "Taking hold of system of systems management," Eng. Manag. J., 2008, to be published.
78. B. Sauser, J. Boardman, and A. Gorod, "SoS management," in System of systems engineering—Innovations for the 21st Century, M. Jamshidi, Ed. Hoboken, NJ: Wiley, 2008.
79. Information Processing Systems—Open Systems Interconnect—Basic Reference Model: Part 4—Management Framework, ISO/IEC 7498, International Standards Organization, 1989.
80. T. Bass, "Managing net-centric systems: The complexity of web services management," presented at the AFEI's Int. Conf. Enterprise Transform., Washington, DC, 2005.
81. S. Dobson, S. Denazis, A. Fernandez, D. Gaiti, E. Gelenbe, F. Massacci, P. Nixon, F. Saffre, N. Schmidt, and F. Zambonelli, "A survey of autonomic communications," ACM Trans. Autonomous Adapt. Syst., vol. 1, no. 2, pp. 223–259, 2006.
82. A. Gorod, B. Sauser, and J. Boardman, "Paradox: Holarchical view of system of systems engineering management," presented at the IEEE 3rd Int. Conf. Syst. Syst. Eng., Monterey, CA, 2008.
83. U.S. Coast Guard, Washington, DC, "Integrated deepwater system fact sheet," 2007.
84. R. O'Rourke, "Coast guard deepwater program: Background, oversight issues, and options for congress," 2007.
85. Integrated Coast Guard Systems, "Deepwater program trifold," 2007.
86. Office of Information Technology, "Improvements needed in the U.S. Coast Guard's acquisition and implementation of deepwater informa-tion technology systems," OIG-06-55, 2006.
87. M. Arnberg, "SoS challenges in USCG deepwater program," presented at the 1st Ann. Syst. Syst. Eng. Conf., Johnstown, PA, 2005.

# *chapter seven*

# *Waterfall model, V-model, spiral model, and other SE models*

## *Introduction*

Several systems engineering (SE) models are currently being practiced (Satzinger, Jackson, and Burd, 2016; Colombi and Cobb, 2009; Furness, 2018; Haas, Perry, and Rephlo, 2009; Henning and Walter, 2012; Jia-Ching, 2011; London, 2012; McMurtry, 2013; DAU, 2017; Tutorials Point, 2018). In many cases, the applications are customized for internal organizational applications and are not fully documented in the open literature. The most common models include the waterfall model, the V-model, the spiral model, the walking skeleton model, and others, many of which originated in the software development industry. Selected ones are described below.

## *The waterfall model*

The waterfall model, also known as the linear-sequential life cycle model, breaks down the SE development process into linear sequential phases that do not overlap one another. The model can be viewed as a flow-down approach to engineering development. As seen in Figures 7.1 and 7.2, the waterfall model assumes that each preceding phase must be completed before the next phase can be initiated. Additionally, each phase is reviewed at the end of its cycle to determine whether or not the project aligns with the project specifications, needs, and requirements. Although the orderly progression of tasks simplifies the development process, the waterfall model is unable to handle incomplete tasks or changes made later in the life cycle without incurring high costs. This makes sense for the waterfall model since water normally flows downward, unless forced to go upward through a pumping device, which could be an additional cost. Therefore, this model lends itself better to simple projects that are well defined and understood.

Case study: This model was instrumental to updating the Conway Regional Medical Center in Arkansas. In 2010, the regional center still did not have an electronic database and information management system for its home health care patients. Consequently, the hospital applied the waterfall method to their acquisition of software to handle their needs.

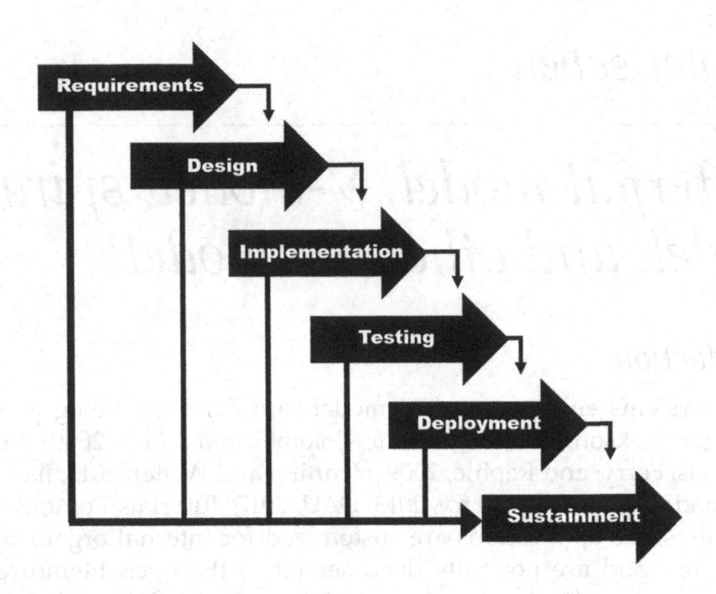


*Figure 7.1* Waterfall model block diagram format.

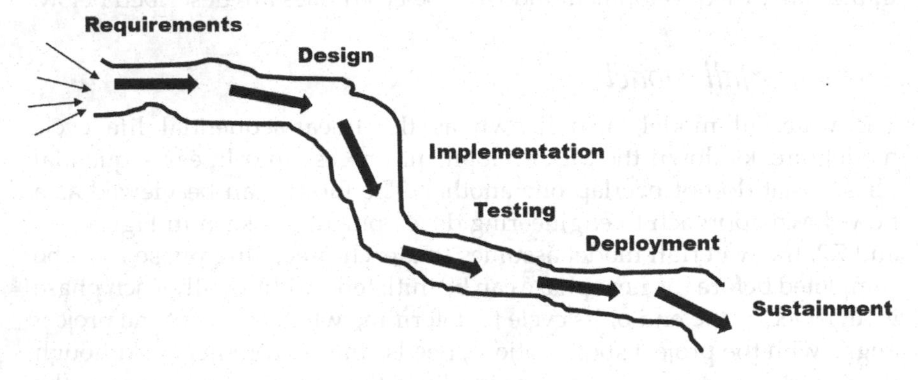


*Figure 7.2* Waterfall model block flow pipe format.

First, the hospital management defined their problem as needing a way to maintain a database of documentation and records for the home health care patients. The hospital then elected to buy, rather than design, the software necessary for this project. After defining their system requirements, the hospital's administration team purchased what they evaluated to be the most suitable software option. However, the hospital performed systems testing before integrating the software into the home health care system. Upon completion of testing, the hospital found that the code needed to be updated once every 6 weeks. This update was factored into their operation and maintenance plan for use of this new system. The system was finally deemed a resounding success once the system was implemented, tested, and the operations and maintenance schedules were created. Through the use of the waterfall model, the software got on track for installation as the primary home health care software for the Conway Regional Medical Center.

## The V-model

The V-model, or the verification and validation model, is an enhanced version of the waterfall model that illustrates the various stages of the system life cycle, as shown in Figures 7.3 and 7.4. The V-model is similar to the waterfall model in that they are both linear models whereby each phase is verified before moving on to the next phase. Beginning from the left side, the V-model depicts the development actions that flow from concept of operations to the integration and verification activities on the right side of the diagram. With this model, each phase of the life cycle has a corresponding test plan that helps identify errors early in the life cycle, minimize future issues, and verify adherence to project specifications. Thus, the V-model lends itself well to proactive defect testing and tracking. However, a drawback of the V-model is that it is rigid and offers

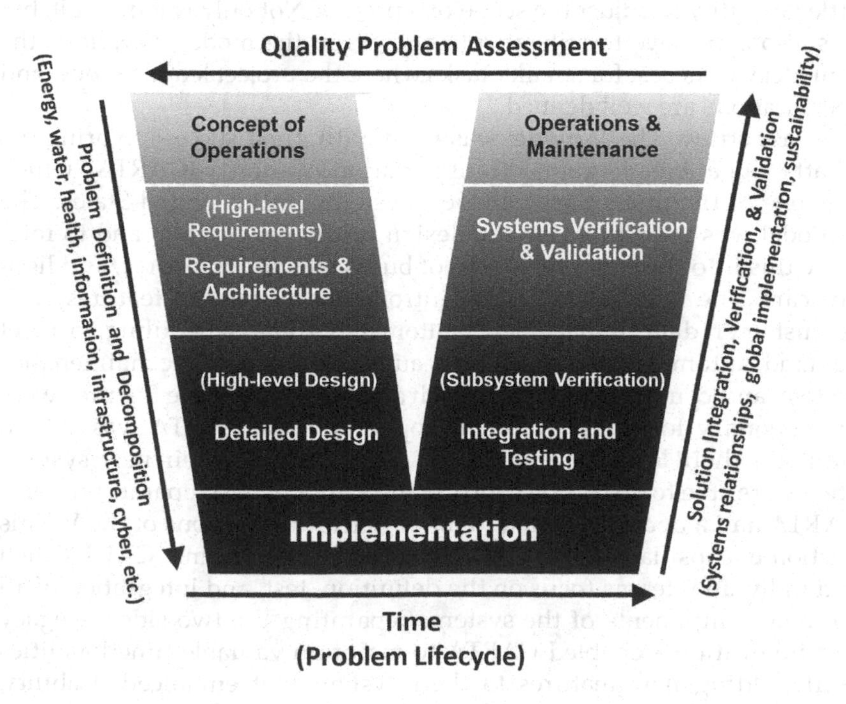


*Figure 7.3* The V-model of SE.

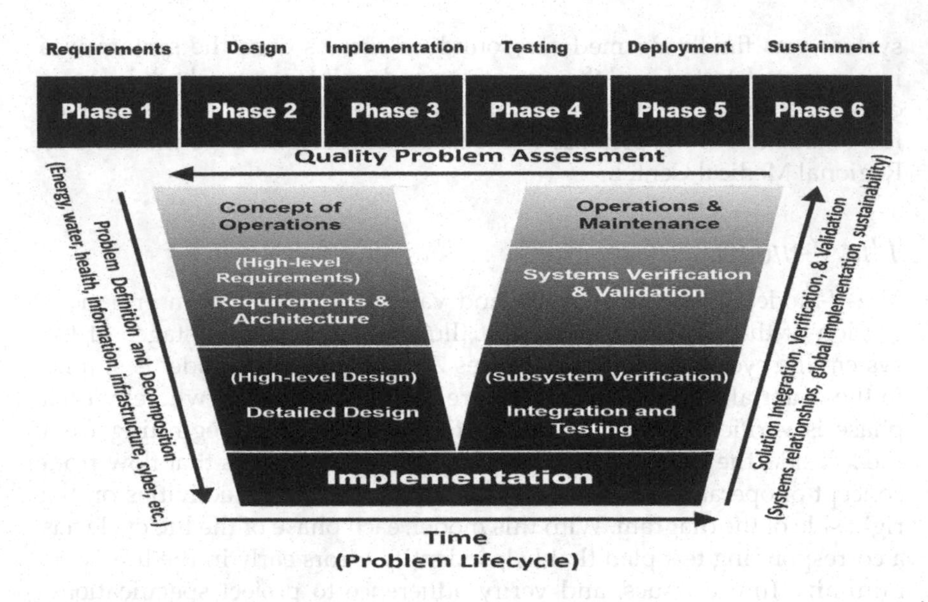


*Figure 7.4* The V-model with phases.

little flexibility to adjust the scope of a project. Not only is it difficult, but it is also expensive to reiterate phases within the model. Therefore, the V-model works best for smaller tasks where the project length, scope, and specifications are well defined.

Case study: The V-model was used with great success to bring the Chattanooga Area Regional Transportation Authority (CARTA), which was one of the first smart transport systems in the United States. The V-model was used to guide the design of the new system and to integrate this into the existing system of buses, electric transport, and light rail cars. The new smart system introduced a litany of features such as customer data management, automated route scheduling to meet demand, automated ticket vending, automated diagnostic maintenance system, and computer aided dispatch and tracking. These features were revolutionary for a mid-sized metropolitan area. CARTA was able to maintain their legacy transport, while integrating their new system. They were able to do this by splitting the V-model into separate sections. CARTA had a dedicated team to manage the flat portions of the V. This portion encapsulated the legacy transportation systems. CARTA then had individual teams focus on the definition, test, and integration of all the new components of the system. Separating the two sides – legacy and innovation – enabled CARTA to maintain valuable functionalities while adding new features to their system that enhanced usability, safety, and efficiency.

## Spiral model

The spiral model is similar to the V-model in that it references many of the same phases as part of its color coordinated slices (shown in Figure 7.5), which indicates the project's stage of development. This model enables multiple flows through the cycle to build a better understanding of the design requirements and engineering complications. The reiterative design lends itself well to supporting popular model based SE techniques such as rapid prototyping and quick failure methods. Additionally, the model assumes that each iteration of the spiral will produce new information that will encourage technology maturation, evaluate the project's financial situation, and justify continuity. Consequently, the lessons learned from this method provide a data point with which to improve the product. Generally, the spiral model meshes well with the defense life cycle management vision and integrates all facets of design, production, and integration.

The spiral model is the foundation of the RQ-4 Global Hawk Operational Management and Usage platform. The Global Hawk was phased into operation in six distinct spirals, with each spiral adding new capabilities to the airframe. The first spiral was getting the aircraft in the sky and having a support network to keep it in the air. Everything from pilots to maintainers was optimized to keep the Global Hawk in the air as much as possible. The subsequent phases added imagery (IMINT), signal (SIGINT), radar, and survivability capabilities to the airframe. Each of

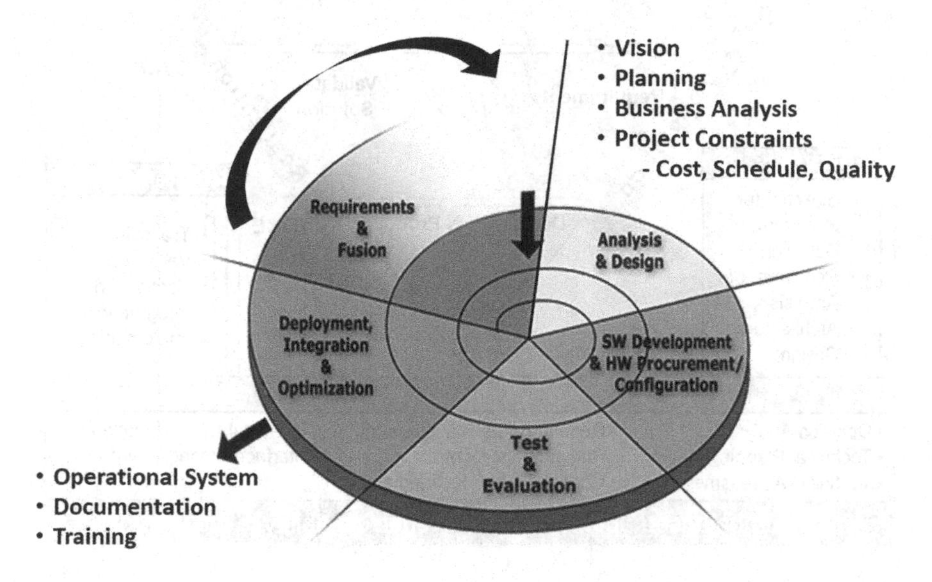


*Figure 7.5* The spiral model of SE.

these capabilities was added one at a time in a spiral development cycle to ensure that each one was integrated into the airframe to the operational standard, and could be adjusted to meet this standard before moving on to the next capability. The benefit of incrementally adding capabilities in a spiral fashion greatly helped the Global Hawk stay on budget and schedule for operational rollout.

## *Defense Acquisitions University SE model*

The new Defense Acquisitions University (DAU) model for SE also originates from the V-model. However, unlike the traditional V-model, it allows for process iteration similarly to the spiral model. As shown in Figure 7.6, a unique attribute of the DAU process is that its life cycle does not need to be completed in order to gain the benefit of iteration. Whereas the spiral model requires the life cycle process to be completed, the DAU model can refine and improve products at any point in its phase progression. This design is beneficial to making early stage improvements, which helps systems engineers to avoid budgeting issues such as cost overruns. Moreover, the model allows for fluid transition between project definition

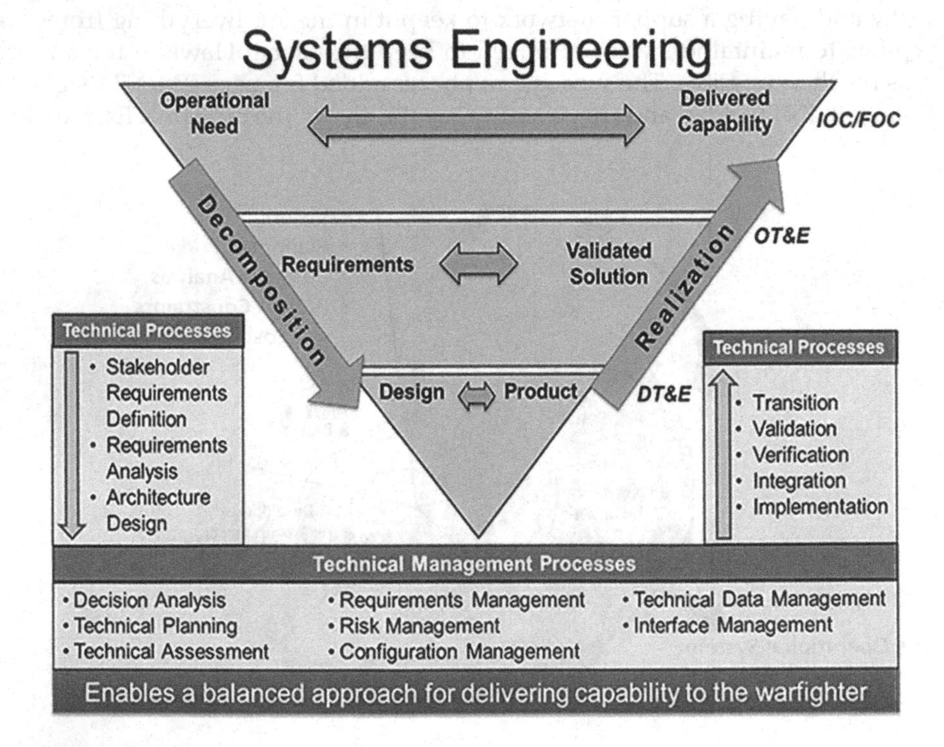


***Figure 7.6*** 2014 edition of DAU SE model.

(decomposition) and product completion (realization), which is useful in software production and integration. Overall, the DAU model is a fluid combination of the V-model and spiral model.

This tailored V-model was used by the Air Force researchers to create a system to aid Battlefield Airmen in identifying friendly forces and calling in close air support with minimal risk to ground troops. The model was used to find an operational need and break it down into a hierarchy of objectives. The researchers used the hierarchy to design multiple prototypes that attempted to incorporate all of the stated objectives. They used rapid prototyping methods to produce these designs, and they were then tested operationally within battlefield airmen squadrons. Ultimately, the production of a friendly marking device was achieved, and this valuable capability was able to be delivered to the warfighter.

## *Walking skeleton model*

The walking skeleton model is a lean approach to incremental development, popularly used in software design. It centers on creating a skeleton framework for what the system is going to do and look like. This basic starting point of the system will have minimal functionality, and the systems engineer will work to add muscles to the skeleton (Figure 7.7). The first step creates a system that may do a very basic yet integral part of the final system design. For example, if one were to design a car using this method, the skeleton would be an engine attached to a chassis with wheels. Once the first basic step is done, the muscles begin to be added to the skeleton. These muscles are more refined and are added one at a time, meaning that each new feature of the system must be completed to add. Furthermore, it is highly recommended that the most difficult features of the system are the first muscles to be added. System components that take a lot of time, require contracting/outsourcing, or are the primary payload must be the first to be completed. This will become the heart of the

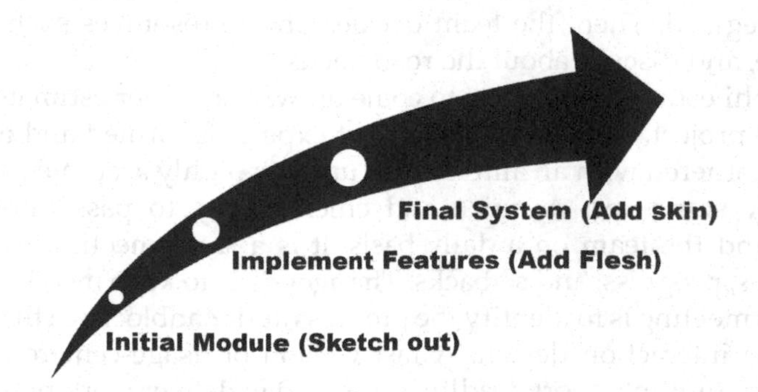


*Figure 7.7* Walking skeleton model.

skeleton, and the rest of the architecture can be optimized to ensure the critical capability of the system is preserved and enhanced.

An example of engineering using the walking skeleton model is the Boston Dynamics walking dogs. The first thing that the engineers did for these robotic creatures was to create a power source and mobility framework. From there, the engineers were able to then go piece by piece and add more functionality to the project, such as the ability to open doors, pick up objects, and even carry heavy loads. The lessons they learned from adding muscles to their skeleton allowed them to move in leaps and bounds, and the benefits were felt across their entire network of products.

Walking skeleton technique varies with the system being developed. In case of a client-server system, it will be a single screen connected for navigating to database and back to screen.

In a front-end application system, it acts as a connection between the platforms and compilation takes place for the simplest element of the language. In a transaction process, it is walking through a single transaction.

Following are the techniques which can be used to create a walking skeleton:

- Methodology shaping: Gathering information about prior experiences and using it to come up with the starter conventions. Following two steps are used in this technique:
  1. Project interviews
  2. Methodology shaping workshop
- Reflection workshop: A particular workshop format for reflective improvement. In the reflection workshop team members discuss what is working fine, what improvements are required and what unique things will be added next time.
- Blitz planning: Every member involved in project planning notes all the tasks on the cards, which will then be sorted, estimated, and strategized. Then, the team decides on the resources such as cost, time, and discuss about the roadblocks.
- Delphi estimation: A way to come up with a starter estimate for the total project. A group consisting of experts is formed and opinions are gathered with an aim to come up with highly accurate estimates.
- Daily stand-ups: A quick and efficient way to pass information around the team on a daily basis. It is a short meeting to discuss status, progress, and setbacks. The agenda is to keep meetings short. This meeting is to identify the progress and roadblocks in the project.
- Agile interaction design: A fast version of usage-centered design, where multiple short deadlines are used to deliver working software without giving important considerations to activities of designing.

To simplify the user–interface test, LEET a record/Capture tool is used.

- Process miniature: A learning technique as any new process is unfamiliar and time consuming. When the process is complex, more time is required for new team members to understand how different parts of the process fit. Time taken to understand the process is reduced with use of process miniature.
- Side-by-side programming: An alternative of pair programming is "Programming in Pairs." Here two people work on one assignment by taking turns in providing input and mostly on a single workstation. It results in better productivity and cost consumed for fixing bugs is less.

  Programmers work without interfering in their individual assignments and review each other's work easily.
- Burn charts: This tool is used to estimate actual and estimated amount of work against the time.

## Object-oriented analysis and design

Object-oriented analysis and design (OOAD) is an Agile Methodology approach toward SE, and eschews traditional systems design processes. Traditional methods demand complete and accurate requirement specification before development; agile methods presume that change is unavoidable and should be embraced throughout the product development cycle. This is a foreign concept to many systems engineers that follow precise documentation habits, and would require an overhaul of project management architecture in order to work. If the necessary support is in place to allow for this approach, it works by grouping data, processes, and components into similar objects. These are the logical components of the system which interact with each other.

For example, customers, suppliers, contracts, and rental agreements would be grouped in a single object together. This object would then be managed by a single person with complete executive control over the data and relationships within. This approach is people based, relying on the individual competencies and exquisite knowledge of their respective object. The systems manager then needs to link all of the people and their objects together to create the final system. The approach hinges upon each person perfecting their object that they are in charge of, and the systems engineer puts all of the pieces together. It puts all of the design control in the hands of the individual engineers. The most popular venue for use of this type of SE is software engineering. It allows experts within their fields to focus on what they do best for a program. OOAD does not allow for convenient system oversight, process verification, or even schedule management, and as such makes it very difficult to get consistent project updates. While it may be conducive to small team projects, this method is unlikely to be feasible for any Air Force project of record.

## References

Colombi, J. and Cobb, R. (October 2009). Application of systems engineering to rapid prototyping for close air support, *Defense Acquisition Research Journal* 16(3), 284–303.

DAU. (29 September 2017). Systems engineering process. *Defense Acquisition University*. www.dau.mil/acquipedia/Pages/ArticleDetails.aspx?aid=9c591ad6-8f69-49dd-a61d-4096e7b3086c.

Furness, D. (12 June 2018). From BigDog to SpotMini: Tracing the evolution of Boston dynamics robo-dogs. *Digital Trends*. www.digitaltrends.com/cool-tech/evolution-boston-dynamics-spot-mini/.

Haas, R., Perry, E., and Rephlo, J. (31 October 2009). A case study on applying the systems engineering approach: Best practices and lessons learned from the Chattanooga SmartBus Project. Home - Transport Research International Documentation – TRID. https://trid.trb.org/view.aspx?id=920390.

Henning, W.A. and Walter, D.T. (December 2012). *Spiral Development in Action: A Case Study of Spiral Development in the Global Hawk Unmanned Aerial Vehicle Program*. Naval Postgraduate School, Monterey, CA. https://calhoun.nps.edu/bitstream/handle/10945/9980/05Dec_Henning_MBA.pdf?sequence=1.

Jia-Ching, L. (11 August 2011). *Various Approaches for Systems Analysis and Design*. Populations and Sampling. University of Missouri, St. Louis. www.umsl.edu/~sauterv/analysis/termpapers/f11/jia.html.

London, B. (8 March 2012). *A Model-Based Systems Engineering Framework for Concept Development*. PDF. MIT Press, Cambridge, MA.

McMurtry, M. (1 January 2013). A case study of the application of the systems development life cycle (SDLC) in 21st century health care: Something old, something new? Papers of the Abraham Lincoln Association, Michigan Publishing, University of Michigan Library. https://quod.lib.umich.edu/j/jsais/11880084.0001.103/--case-study-of-the-application-of-the-systems-development?rgn=main%3Bview.

Satzinger, J.W., Jackson, R.B., and Burd, S.D. (2016). *Approaches to Systems Development. In Systems Analysis and Design in a Changing World*, Sixth Edition. Cengage Learning, Independence, KY. pp. 8-1–8-9.

Tutorials Point. (21 July 2018). V-Model. Retrieved on 08/08/18 at www.tutorialspoint.com/sdlc/sdlc_v_model.htm.

# chapter eight

# The DEJI model of systems engineering

## Introduction

Systems quality is at the intersection of efficiency, effectiveness, and productivity. Efficiency provides the framework for quality in terms of resources and inputs required to achieve the desired level of quality. Effectiveness comes into play with respect to the application of product quality to meet specific needs and requirements of an organization. Productivity is an essential factor in the pursuit of quality as it relates to the throughput of a production system. To achieve the desired levels of quality, efficiency, effectiveness, and productivity, a new research framework must be adopted. In this chapter, we present a quality enhancement model for quality Design, Evaluation, Justification, and Integration (DEJI) based on a product development application. The model is relevant for research efforts in quality engineering and technology applications and other systems engineering applications.

The DEJI model of systems engineering provides one additional option for systems engineering development applications. Although the model is generally applicable in all types of systems modeling, systems quality is specifically used to describe how the DEJI model is applied. The core stages of the DEJI model are:

- Design
- Evaluation
- Justification
- Integration.

Design encompasses any system initiative providing a starting point for a project. Thus, design can include technical product design, process initiation, and concept development. In essence, we can say that "design" represents requirements and specifications. Evaluation can use a variety of metrics both qualitative and quantitative, depending on the organization's needs. Justification can be done on the basis of monetary, technical, or social reasons. Integration needs to be done with respect to the normal or standard operations of the organization. Figure 8.1 illustrates the full profile of the DEJI model.

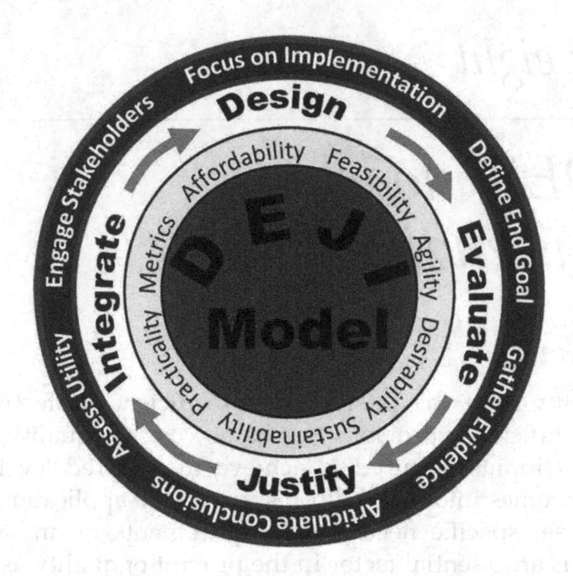


***Figure 8.1*** The DEJI model (see www.DEJImodel.com for model details).

All the operational elements embedded in the DEJI model are explained and described as presented below:

Design embodies Agility, Define End Goal, and Engage Stakeholder.
Evaluate embodies Feasibility, Metrics, Gather Evidence, and Assess Utility.
Justify embodies Desirability, Focus on Implementation, and Articulate Conclusions.
Integrate embodies Affordability, Sustainability, and Practicality.
For application purposes, these elements interface and interact systematically to enhance overall operational performance of an organization.

## *Application to quality system*

Several aspects of quality must undergo rigorous research along the realms of both quantitative and qualitative characteristics. Many times, quality is taken for granted and the flaws only come out during the implementation stage, which may be too late to rectify. The growing trend in product recalls is a symptom of a priori analysis of the sources and implications of quality at the product conception stage. This column advocates the use of the DEJI model for enhancing quality design, quality evaluation, quality justification, and quality integration through hierarchical and stage-by-stage processes.

Better quality is achievable, and there is always room for improvement in the quality of products and services. But we must commit more efforts to the research at the outset of the product development cycle. Even the human elements of the perception of quality can benefit from more directed research from a social and behavioral sciences point of view.

## *Quality accountability*

Throughout history, engineering has answered the call of the society to address specific challenges (Grayson, 1993; Kirby et al., 1990). With such answers comes a greater expectation of professional accountability. Consider the level of social responsibility that existed during the time of the Code of Hammurabi (Bryant, 2005). Two of the laws are echoed below:

> **Hammurabi's Law 229**
>
> If a builder build a house for someone, and does not construct it properly, and the house which he built fall in and kill its owner, then that builder shall be put to death.
>
> **Hammurabi's Law 230**
>
> If it kills the son of the owner the son of that builder shall be put to death.

These are drastic measures designed to curb professional dereliction of duty and enforce social responsibility with particular focus on product quality. Research and education must play bigger and more direct roles in the design, practice, and management of quality. Duckworth and Moore (2010) present modern aspects of social responsibility in the context of day-to-day personal and professional activities. The global responsibility of the greater society is presented by Bhargava (2006) with respect to world development challenges covering the global economy, human development, global governance, and social relationships. Quality is the common theme in the development challenges. Focusing on the emerging field of Big Data, Xian and Madhavan (2014) advocate engineering education collaboration, which aligns well with data-intensive product development. With the above principles as possible tenets for better research, education, and practice of quality in engineering and technology, this chapter presents the DEJI model as a potential methodology. Pertinent considerations are provided by Badiru (2013) and Miller (2010).

## The DEJI model

This model encourages the practice of building quality into a product right from the beginning so that the product integration stage can be more successful. Figure 8.2 shows the four-legged model for an overall systems quality with respect to design, evaluation, justification, and integration. Table 8.1 shows the potential elements within each stage of the DEJI model.

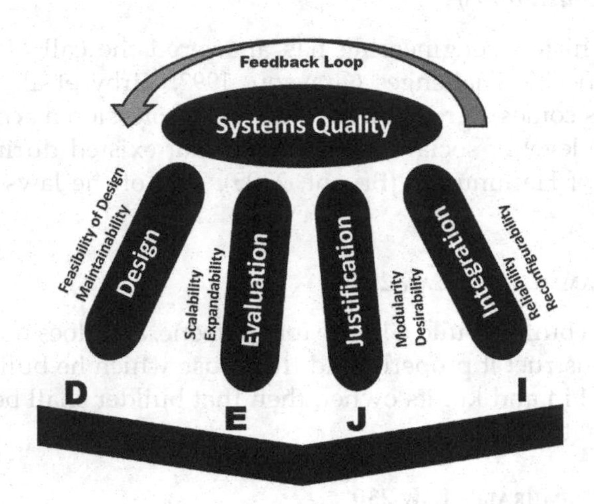


*Figure 8.2* The DEJI model of design, evaluation, justification, and integration.

*Table 8.1* Tabulated layout of the DEJI model

| DEJI model | Characteristics | Tools & techniques |
|---|---|---|
| Design | Define goals | Parametric assessment |
| | Set performance metrics | Project state transition |
| | Identify milestones | Value stream analysis |
| Evaluate | Measure parameters | Pareto distribution |
| | Assess attributes | Life cycle analysis |
| | Benchmark results | Risk assessment |
| Justify | Assess economics | Benefit–cost ratio |
| | Assess technical output | Payback period |
| | Align with goals | Present value |
| Integrate | Embed in normal operation | SMART concept |
| | Verify symbiosis | Process improvement |
| | Leverage synergy | Quality control |

## Design of quality

The design of quality in product development should be structured to follow point-to-point transformations. A good technique to accomplish this is the use of state-space transformation, with which we can track the evolution of a product from the concept stage to a final product stage. For the purpose of product quality design, the following definitions are applicable:

**Product state**: A state is a set of conditions that describe the product at a specified point in time. The *state* of a product refers to a performance characteristic of the product which relates input to output such that a knowledge of the input function over time and the state of the product at time $t = t_0$ determines the expected output for $t \geq t_0$. This is particularly important for assessing where the product stands in the context of new technological developments and the prevailing operating environment.

**Product state-space**: A product *state-space* is the set of all possible states of the product lifecycle. State-space representation can solve product design problems by moving from an initial state to another state, and eventually to the desired end-goal state. The movement from state to state is achieved by means of actions. A goal is a description of an intended state that has not yet been achieved. The process of solving a product problem involves finding a sequence of actions that represents a solution path from the initial state to the goal state. A state-space model consists of state variables that describe the prevailing condition of the product. The state variables are related to inputs by mathematical relationships. Examples of potential product state variables include schedule, output quality, cost, due date, resource, resource utilization, operational efficiency, productivity throughput, and technology alignment. For a product described by a system of components, the state-space representation can follow the quantitative metric below:

$$Z = f(z,x); \quad Y = g(z,x)$$

where $f$ and $g$ are vector-valued functions. The variable $Y$ is the output vector while the variable $x$ denotes the inputs. The state vector $Z$ is an intermediate vector relating $x$ to $y$. In generic terms, a product is transformed from one state to another by a driving function that produces a transitional relationship given by

$$S_s = f(x \mid S_p) + e$$

where $S_s$ = subsequent state; $x$ = state variable; $S_p$ = the preceding state; and $e$ = error component.

The function *f* is composed of a given action (or a set of actions) applied to the product. Each intermediate state may represent a significant milestone in the project. Thus, a descriptive state-space model facilitates an analysis of what actions to apply in order to achieve the next desired product state. Figure 8.3 shows a representation of a product development example involving the transformation of a product from one state to another through the application of human or machine actions. This simple representation can be expanded to cover several components within the product information framework. Hierarchical linking of product elements provides an expanded transformation structure. The product state can be expanded in accordance with implicit requirements. These requirements might include grouping of design elements, linking precedence requirements (both technical and procedural), adapting to new technology developments, following required communication links, and accomplishing reporting requirements. The actions to be taken at each state depend on the prevailing product conditions. The nature of subsequent alternate states depends on what actions are implemented. Sometimes there are multiple paths that can lead to the

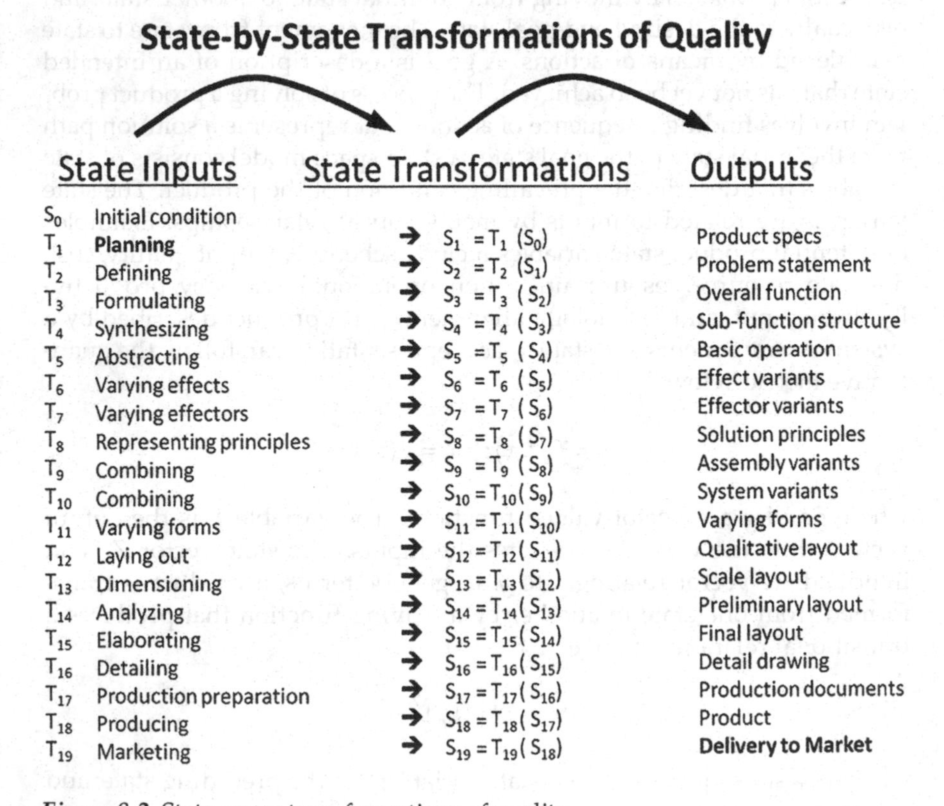

State-by-State Transformations of Quality
State Inputs
State Transformations
Outputs
S0 Initial condition
T1 Planning → S1 = T1 (S0) Product specs
T2 Defining → S2 = T2 (S1) Problem statement
T3 Formulating → S3 = T3 (S2) Overall function
T4 Synthesizing → S4 = T4 (S3) Sub-function structure
T5 Abstracting → S5 = T5 (S4) Basic operation
T6 Varying effects → S6 = T6 (S5) Effect variants
T7 Varying effectors → S7 = T7 (S6) Effector variants
T8 Representing principles → S8 = T8 (S7) Solution principles
T9 Combining → S9 = T9 (S8) Assembly variants
T10 Combining → S10 = T10 (S9) System variants
T11 Varying forms → S11 = T11 (S10) Varying forms
T12 Laying out → S12 = T12 (S11) Qualitative layout
T13 Dimensioning → S13 = T13 (S12) Scale layout
T14 Analyzing → S14 = T14 (S13) Preliminary layout
T15 Elaborating → S15 = T15 (S14) Final layout
T16 Detailing → S16 = T16 (S15) Detail drawing
T17 Production preparation → S17 = T17 (S16) Production documents
T18 Producing → S18 = T18 (S17) Product
T19 Marketing → S19 = T19 (S18) Delivery to Market


*Figure 8.3* State-space transformations of quality.

desired end result. At other times, there exists only one unique path to the desired objective. In conventional practice, the characteristics of the future states can only be recognized after the fact, thus making it impossible to develop adaptive plans. In the implementation of the DEJI model, adaptive plans can be achieved because the events occurring within and outside the product state boundaries can be taken into account.

If we describe a product by $P$ state variables $s_i$, then the composite state of the product at any given time can be represented by a vector **S** containing $P$ elements. That is,

$$\boldsymbol{S} = \{s_1, s_2, \ldots, s_P\}$$

The components of the state vector could represent either quantitative or qualitative variables (e.g., cost, energy, color, time). We can visualize every state vector as a point in the state-space of the product. The representation is unique since every state vector corresponds to one and only one point in the state-space. Suppose we have a set of actions (transformation agents) that we can apply to the product information so as to change it from one state to another within the project state-space. The transformation will change a state vector into another state vector. A transformation may be a change in raw material or a change in design approach. The number of transformations available for a product characteristic may be finite or unlimited. We can construct trajectories that describe the potential states of a product evolution as we apply successive transformations with respect to technology forecasts. Each transformation may be repeated as many times as needed. Given an initial state $\boldsymbol{S}_0$, the sequence of state vectors is represented by the following:

$$\boldsymbol{S}_n = T_n(\boldsymbol{S}_{n-1}).$$

The state-by-state transformations are then represented as $\boldsymbol{S}_1 = T_1(\boldsymbol{S}_0)$; $\boldsymbol{S}_2 = T_2(\boldsymbol{S}_1); \boldsymbol{S}_3 = T_3(\boldsymbol{S}_2); \ldots; \boldsymbol{S}_n = T_n(\boldsymbol{S}_{n-1})$. The final State, $\boldsymbol{S}_n$, depends on the initial state **S** and the effects of the actions applied.

## *Evaluation of quality*

A product can be evaluated on the basis of cost, quality, schedule, and meeting requirements. There are many quantitative metrics that can be used in evaluating a product at this stage. Learning curve productivity is one relevant technique that can be used because it offers an evaluation basis of a product with respect to the concept of growth and decay. The half-life extension (Badiru, 2012) of the basic learning is directly applicable because the half-life of the technologies going into a product can be considered. In today's technology-based operations, retention of learning may be threatened by fast-paced shifts in operating requirements. Thus, it is of interest to evaluate the half-life properties of new technologies as they impact the overall product quality. Information about the half-life can tell us something about the sustainability of learning-induced technology performance. This is particularly useful for designing products whose life cycles stretch into the future in a high-tech environment.

## *Justification of quality*

We need to justify a program on the basis of quantitative value assessment. The Systems Value Model (SVM) is a good quantitative technique that can be used here for project justification on the basis of value. The model provides a heuristic decision aid for comparing project alternatives. It is presented here again for the present context. Value is represented as a deterministic vector function that indicates the value of tangible and intangible attributes that characterize the project. It is represented as

$$V = f(A_1, A_2, \ldots, A_p)$$

where $V$ is the assessed value and the $A$ values are quantitative measures or attributes. Examples of product attributes are quality, throughput, manufacturability, capability, modularity, reliability, interchangeability, efficiency, and cost performance. Attributes are considered to be a combined function of factors. Examples of product factors are market share, flexibility, user acceptance, capacity utilization, safety, and design functionality. Factors are themselves considered to be composed of indicators. Examples of indicators are debt ratio, acquisition volume, product responsiveness, substitutability, lead time, learning curve, and scrap volume. By combining the above definitions, a composite measure of the operational value of a product can be quantitatively assessed. In addition to the quantifiable factors, attributes, and indicators that impinge upon overall project value, the human-based subtle factors should also be included in assessing overall project value.

## *Earned value technique for earned quality*

Value is synonymous with quality. Thus, the contemporary earned value technique is relevant for "earned quality" analysis. This is a good analytical technique to use for the justification stage of the DEJI model. This will impact cost, quality, and schedule elements of product development with respect to value creation. The technique involves developing important diagnostic values for each schedule activity, work package, or control element as shown in Figure 8.4. The variables in the figure are as follows: PV: Planned Value; EV: Earned Value; AC: Actual Cost; CV: Cost Variance; SV: Schedule Variance; EAC: Estimate at Completion; BAC: Budget at

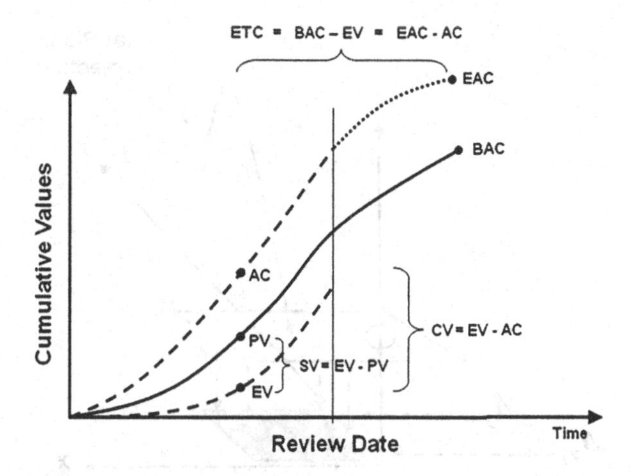


*Figure 8.4* Earned value technique for quality assessment.

Completion; and ETC: Estimate to Complete. This analogical relationship is a variable research topic for quality engineering and technology applications.

## *Integration of quality*

Without being integrated, a system will be in isolation and it may be worthless. We must integrate all the elements of a system on the basis of alignment of functional goals. The overlap of systems for integration purposes can conceptually be viewed as projection integrals by considering areas bounded by the common elements of subsystems as shown in Figure 8.5. Quantitative metrics can be applied at this stage for effective assessment of the product state. Trade-off analysis is essential in quality integration. Pertinent questions include the following:

What level of trade-offs on the level of quality are tolerable?
What is the incremental cost of higher quality?
What is the marginal value of higher quality?
What is the adverse impact of a decrease in quality?
What is the integration of quality of time? In this respect, an integral of the form below may be suitable for further research:

$$I = \int_{t_1}^{t_2} f(q)\,dq$$

where $I$ = integrated value of quality, $f(q)$ = functional definition of quality, $t_1$ = initial time, and $t_2$ = final time within the planning horizon.

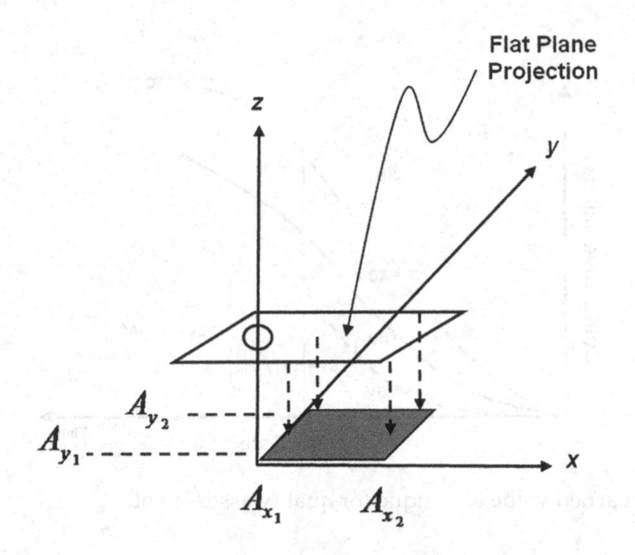


*Figure 8.5* Product quality integration surfaces.

Presented below are guidelines and important questions relevant for quality integration.

- What are the unique characteristics of each component in the integrated system?
- How do the characteristics complement one another?
- What physical interfaces exist among the components?
- What data/information interfaces exist among the components?
- What ideological differences exist among the components?
- What are the data flow requirements for the components?
- What internal and external factors are expected to influence the integrated system?
- What are the relative priorities assigned to each component of the integrated system?
- What are the strengths and weaknesses of the integrated system?
- What resources are needed to keep the integrated system operating satisfactorily?
- Which organizational unit has primary responsibility for the integrated system?

The systems approach of the DEJI model will facilitate a better alignment of product technology with future development and needs. The stages of the model require research for each new product with respect to design, evaluation, justification, and integration. Existing analytical tools and techniques as well as other systems engineering models can be used at each stage of the model. Thus, a hybrid systems modeling is possible.

## Conclusion

Quality is an integrative process that must be evaluated on a stage-by-stage approach. This requires research, education, and implementation strategies that consider several pertinent factors. This column suggests the DEJI model, which has been used successfully for product development applications, as a viable methodology for quality design, quality evaluation, quality justification, and quality integration. This column is intended as a source to spark the interest of researchers to apply this tool in new product development efforts.

## References

Badiru, A.B. (Fall 2012). Application of the DEJI model for aerospace product integration, *Journal of Aviation and Aerospace Perspectives (JAAP)* 2(2), 20–34.

Badiru, A.B. (November 2013). Up to the task: Industrial engineering is essential to meeting NAE's grand challenges, *Industrial Engineer* 45(11), 42–45.

Bhargava, V. (Editor). (2006). *Global Issues for Global Citizens: An Introduction to Key Development Challenges*. The World Bank, Washington, DC.

Bryant, T. (2005). *The Life & Times of Hammurabi*. Mitchell Lane Publishers, Newark, DE.

Duckworth, H.A. and Moore, R.A. (2010). *Social Responsibility: Failure Mode Effects and Analysis*. CRC Press/Taylor & Francis Group, Boca Raton, FL.

Grayson, L.P. (1993). *The Making of an Engineer: An Illustrated History of Engineering Education in the United States and Canada*. John Wiley, New York.

Kirby, R.S., Withington, S., Darlin, A.B., and Kilgour, F.G. (1990). *Engineering in History*. Dover Publications, New York.

Miller, R.K. (19–21 October 2010). The Educational Imperatives of the Engineering Grand Challenges. Keynote Address, ASEE Global Colloquium on Engineering Education. Singapore.

Xian, H. and Madhavan, K. (2014). Anatomy of scholarly collaboration in engineering education: A big-data bibliometric analysis, *Journal of Engineering Education*, 103(3), 486–514.

# chapter nine

# Vehicle systems modeling[1]

## Introduction

Demands for continuous improvements in cost, development cycle time, and customer satisfaction are consistent themes in the global automotive industry. The field of vehicle dynamics is not immune to these pressures. Meeting these demands requires an ever-increasing emphasis on computer simulation and a more comprehensive understanding and focus on customer requirements. Development of vehicle dynamics performance objective tests and metrics is a prerequisite to support these goals.

Standardized tests describing various aspects of vehicle dynamics behavior have been well defined for decades. The majority of these tests focus on delivering metrics that describe steady-state handling characteristics. In contrast, metrics describing transient handling properties are less well defined. Subjective assessment of hardware by experienced professionals is the primary method of vehicle development for transient handling characteristics. To fully understand customer requirements, it is necessary to develop tests and metrics that capture transient attributes.

The goal of this paper is to describe subjective methods for evaluation of transient handling and describe a newly-developed transient handling objective test. The first section categorizes transient handling into four traits. The second section discusses subjective evaluation methods for the above-mentioned four handling categories. The third section describes the limitations of established objective transient tests and illustrates the DFSS method used to develop a steering robot driven transient (RDT) handling test. The final section provides example RDT data from road and simulation tests to show how the specifications of chassis parameters can affect transient handling characteristics.

## Transient handling characteristics

The fundamental steady-state metrics of Understeer, Cornering Compliance, Roll Gradient, Maximum Lateral Acceleration, and Steering

[1] Reprinted and adapted with Permission from Badiru, Ibrahim A. and M. W. Neal (2013), "Use of DFSS Principles to Develop an Objective Method to Assess Transient Vehicle Dynamics," ***SAE International***, Paper No. 2013-01-0708; doi:10.4271/2013-01-0708

Sensitivity are the starting point for the synthesis of any professional chassis design. These concepts have been well understood going back to the work of Maurice Olley in the 1930's [1] and are still the starting point of any formal education in vehicle dynamics. Borrowing from linear control system theory, tests and metrics for transient handling performance quickly followed the development of steady-state metrics. The traditional transient tests are the step steer and swept sine (frequency) vehicle response. These tests provide classical metrics such as response time, settling time, damping ratio, etc. Section 4 discusses some of the limitations of frequency and step steer tests.

Before proceeding it is critical to briefly review the vehicle dynamics concept of Sideslip, see Figure 9.1. Sideslip is the angular difference between a reference axis and the vehicle velocity vector at a specified point on the body. Sideslip is generally referenced relative to steering angle for the front axle, $\beta f$; or relative to vehicle centerline for the rear axle and center of gravity, $\beta r$ and $\beta cg$ respectively. Sideslip angle and sideslip rate are key driver cues for control of the vehicle and are primary factors in the subjective perception of transient performance [2].

At General Motors subjective evaluations of transient handling can be grouped into 4 categories: (1) agility, (2) stability, (3) precision, and (4) initial roll support. These characteristics are described in the following sections.

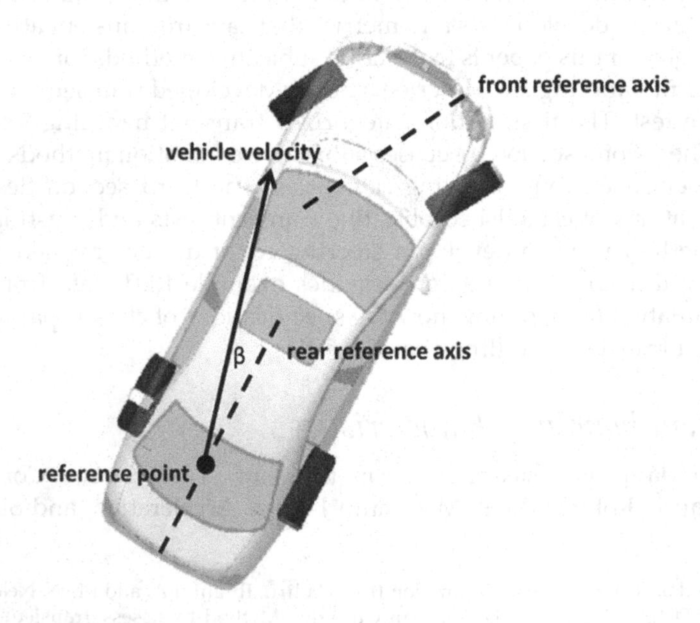


*Figure 9.1* Sideslip at rear axle reference point.

## *Transient agility*

Transient agility is the ability of the vehicle to change heading angle quickly in response to steering input. In a control systems sense this relates to phase lag. A vehicle that responds quickly to steering input provides three advantages to the driver. First, controllability of the system is improved. The driver is better able to predict the exact path the vehicle will follow based on current steering input. Second, the driver perceives the vehicle as more responsive to steering input—a positive attribute for most drivers [2, 3]. Third, the vehicle is better able to quickly maneuver laterally in a short road distance. This is an important trait for obstacle avoidance in emergency situations.

Objectively, agility is related to front axle sideslip, βf in response to a steering input. Transient front axle sideslip is a result of steady-state sideslip characteristics and lag between steering angle change and vehicle heading angle. To illustrate, Figure 9.2 shows a simplified and exaggerated step through the time sequence of a step steering input.

At time step one the vehicle travels forward with wheels straight. At time step two a step steering input has just been completed, but the vehicle has not begun to respond yet. βf is at a peak due to large deviation between road wheel steer angle (RWA) and the vehicle velocity vector (i.e. poor agility). At the final time step βf has reached its steady-state value.

## *Transient stability*

Transient stability is the ability of the vehicle to minimize peak rear axle sideslip, βr and sideslip velocity during a maneuver. Transient rear axle sideslip is a combination of steady-state βr characteristics and overshoot. Evaluators often refer to vehicles with low βr during transient maneuvers as "planted" or "stable". Some evaluators refer to vehicles with high βr as having "Transient Oversteer" [4].

Rear axle sideslip is a direct indicator to the driver of the phasing between yaw velocity and lateral acceleration response. Mathematically there is a direct relationship between yaw velocity and lateral acceleration at steady-state given by:

$$Ay = u \times r \tag{9.1}$$

Where $u$ = forward velocity (m/s), $r$ is yaw velocity (rad/s), and Ay = lateral acceleration (m/s$^2$). While drivers certainly don't perform scientific calculations in their head real time, the culmination of past driving experiences gives people the instinct to detect when yaw velocity exceeds the expected value based on vehicle speed and lateral acceleration experienced. This excessive yaw velocity gives the driver the sense they are not in absolute

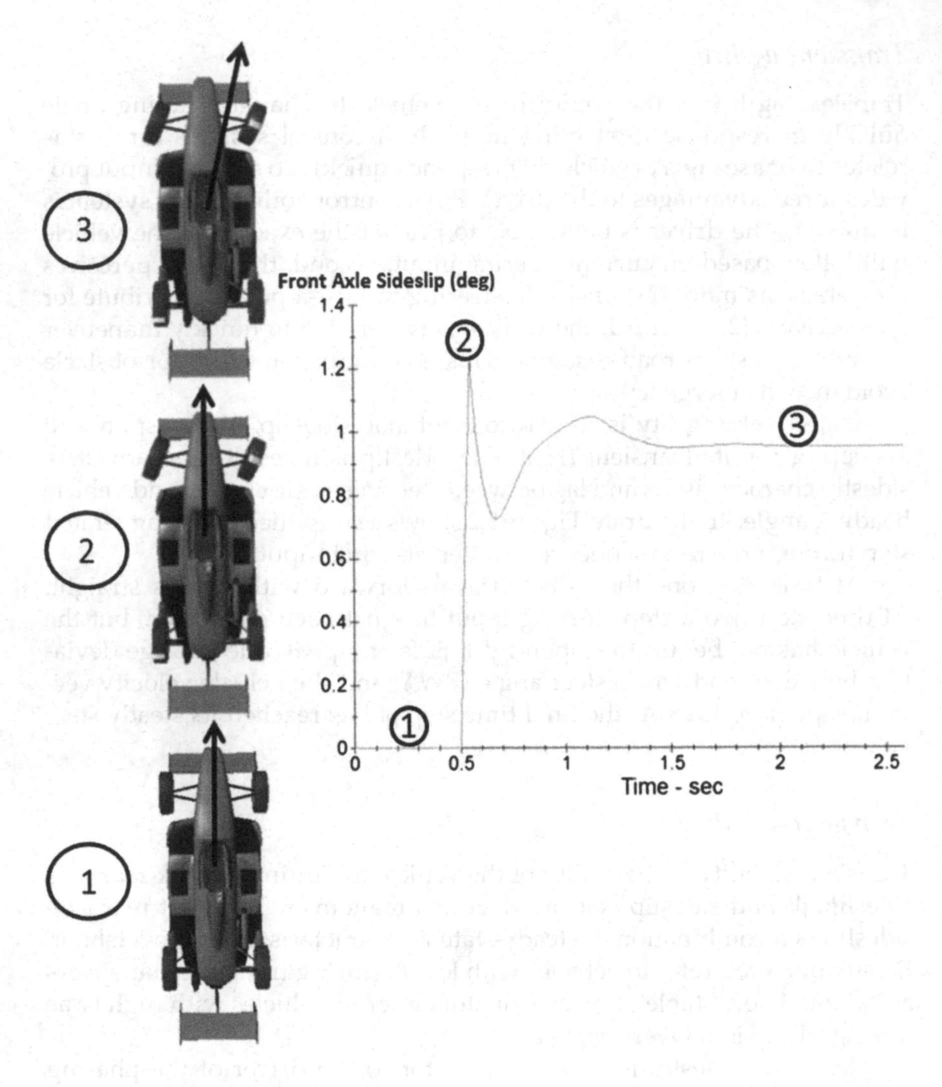


*Figure 9.2* βF vs. time.

control of the vehicle. In fact racecar drivers use this instinct as a primary means to detect approaching loss of directional control when cornering at high lateral accelerations.

We can examine this point from a mathematical point of view by reviewing the generalized equation for *Ay*:

$$Ay = u \times r + u \times \frac{d\beta cg}{dt} \tag{9.2}$$

Where $d\beta/dt$ is sideslip velocity at the cg, rearranging…

$$\frac{d\beta cg}{dt} = \frac{Ay}{u} - r \tag{9.3}$$

In words, $d\beta cg/dt$ is simply the difference between expected steady-state yaw velocity ($Ay/u$) and actual transient yaw velocity ($r$).

### *Transient precision*

We describe the combination of the two traits agility and stability as precision. A vehicle that is precise has low front & rear axle peak side-slip during transient maneuvers. This implies the vehicle yaws quickly in response to steering input and that yaw and lateral acceleration are closely phased. Both of these attributes improve driver confidence to accurately predict vehicle response to a given steering input.

### *Transient roll support*

The final transient performance category quantifies roll response characteristics. Roll support is directly related to vehicle roll velocity at turn initiation. Typically roll gradient is the metric used to communicate roll performance (steady-state metric), but studies show drivers are sensitive to both roll angle and roll velocity [5]. During the initial stages of a steering maneuver roll velocity is the primary cue to the driver about roll behavior. Large roll velocities may give the driver a feeling of impending rollover. The dampers are the primary means of affecting roll velocity. Increased damping forces reduce initial roll velocity.

## *Methods to subjectively evaluate transient handling*

In contrast to the development of formalized analysis and test techniques for handling performance has been the development of subjective assessment techniques. Methods for assessing performance are often ad hoc and there is very little published literature on the subject [3,6,7]. Frequently the closed loop tests of single and double lane change are employed. Closed loop maneuvers have the disadvantage that evaluations can vary considerably amongst different drivers [3].

An alternative to closed loop evaluation is to perform assessments using open-loop inputs. The authors propose the previously defined traits of agility, stability, precision, and roll support can be subjectively evaluated using three open-loop steering maneuvers. First, a ramped step steer input; second, a low amplitude triangle pulse; third, a high amplitude triangle pulse. These steering inputs are described in the next three sections.

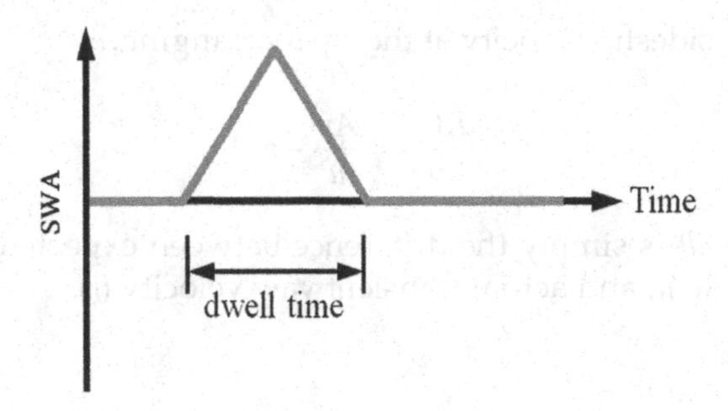


*Figure 9.3* Triangle pulse.

## *Low amplitude triangle pulse steer*

A triangle pulse steer (Figure 9.3) simulates a discrete change in vehicle heading angle and is primarily used to evaluate agility. The evaluator is assessing how accurately vehicle response mimics the steering input. Does the vehicle respond quickly in yaw? Does vehicle heading angle track steering input? install.

## *High amplitude triangle pulse steer*

A high amplitude triangle pulse steer differs from a low amplitude maneuver only in the amplitude of the steering angle and the length of dwell time in the steering event. A larger steering magnitude allows evaluation of stability by building larger lateral accelerations. The evaluator is assessing how well the rear axle tracks with the front. Does the vehicle heading angle match its longitudinal axis during the maneuver? Does vehicle heading angle track steering input?

## *Ramped step steer*

A ramped step steer (Figure 9.4) is similar to a classical step input, except the steering is input over a defined period of time. This input simulates entering a corner. The final steering angle and steering rate can be adjusted to control the peak lateral acceleration.

Evaluators assess agility, stability, precision, and transient roll support using this input for a variety of speeds and lateral accelerations. These characteristics are not only evaluated individually, but evaluated for harmony between the responses. For example, it is important that the phasing of roll and yaw response are correct. Ideally roll velocity should lag lateral response. This ensures the driver does not need to compensate their steering input for roll angle effects.

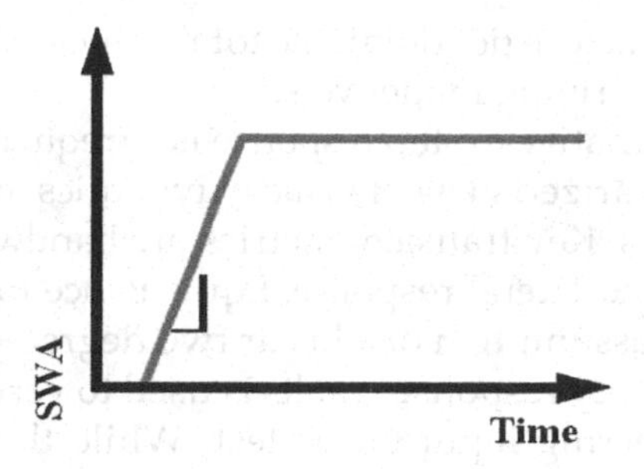


*Figure 9.4* Ramped step steer.

## Development of robot driven transient test

### Limitations of current tests

Section one briefly introduced the step steer and frequency response objective tests for transient metrics. The metrics derived from these tests provide insight into the behavior of vehicles, but do not always fully correlate with subjective observations of vehicle behavior with realistic steering inputs. The reasons for the shortcomings are described in the next two paragraphs.

The primary characteristic of a step function is the rapid velocity of the input. This rapid input may give misleading results when applied to vehicles for two reasons: (1) A step input has significant high frequency content, while the lateral & roll response bandwidth of most vehicles is below 3 Hz—most driving is below 2 Hz. (2) A step input results in an unnatural discontinuity in tire slip angle, αF during the maneuver, see simulation data in Figure 9.5. These two factors mean tire and damper

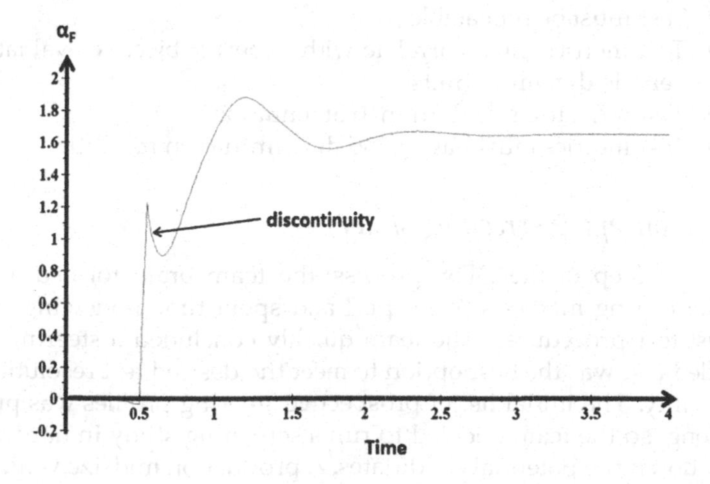


*Figure 9.5* Front tire slip angle with step steer input.

high frequency characteristics dominate total vehicle response to a greater degree than typical driving maneuvers.

The typical alternative to step response is a frequency response input. This test takes a linearized view of vehicle dynamics for the calculation of performance metrics. Key transient metrics are bandwidth and damping ratio for roll, yaw, and lateral response. Equivalence to time domain metrics is made by the assumption of a linear two degree-of-freedom system. Coherence of frequency response results is used to verify this assumption for the range of steering input of the test. While this process provides good results, some of the nuances of performance are lost in the necessary assumption of a linear system.

## *Test development using DFSS*

Design for Six Sigma (DFSS) is a well-established approach to developing products that fulfill customer needs with high performance and robustness. The development team decided the DFSS method could prove advantageous in developing a transient test procedure that correlates with subjective evaluation. For readers unfamiliar with DFSS, a review of the key steps follows:

- Define requirements
- Develop concept
- Optimize design through designed experiment
- Verify final design
- Define Requirements
- A few critical requirements needed to be met as the team began the project:
  - Test must be repeatable.
  - Test metrics must correlate with expert subjective evaluation of vehicle dynamics traits.
  - Test must use standard instrumentation.
  - Test metrics must have good discrimination resolution.

### *Develop concept & screening study*

In the next step of the DFSS process, the team brainstormed a list of potential testing methods "concepts" and spent time reviewing current and past test procedures. The team quickly concluded a steering robot-controlled test was the best option to meet the desired test resolution and repeatability. The initial list of prospective steering profiles was prohibitively long, so the team decided to run a screening study in hardware to narrow down the potential candidates. A production midsize vehicle was instrumented with a data acquisition system capable of measuring vehicle sideslip, lateral acceleration, steering angle, roll angle, roll velocity, and yaw velocity. An Anthony Best robotic steering controller provided steering control.

The vehicle was setup in two chassis configurations. By design, the two chassis configurations had vastly different handling characteristics. Internal experts performed subjective evaluations for both configurations followed by objective testing using six candidate test procedures:

- Fixed frequency single cycle sine, Hz #1
- Fixed frequency single cycle sine, Hz #2
- High amplitude triangle pulse with dwell
- Low amplitude triangle pulse with dwell
- High amplitude ramped step input
- Low amplitude ramped step input

Analysis of the raw data results allowed the team to identify the steering maneuvers and data channels that provided the best chassis performance discrimination. Based on this study step steer and low amplitude triangle pulse were not pursued any further.

## *Optimize design*

The goal of the optimization process is to establish the combination of steering input and objective metrics that best discriminates attribute differences between cars as observed by experts.

## *P-Diagram & ideal function*

Figure 9.6 gives an overview of the system "p-diagram". The p-diagram is a graphical representation of the inputs and outputs of the designed

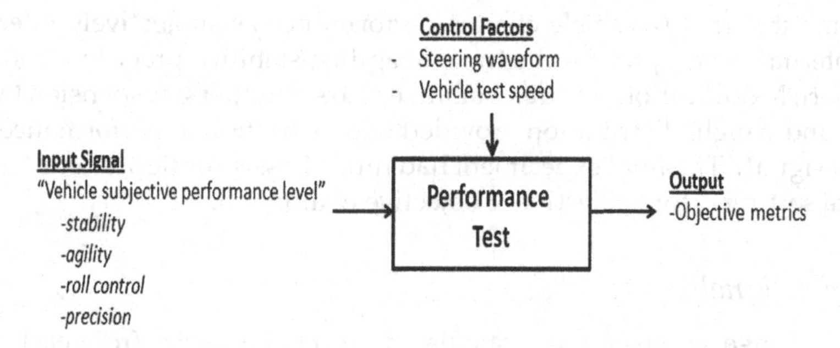


*Figure 9.6* P-diagram.

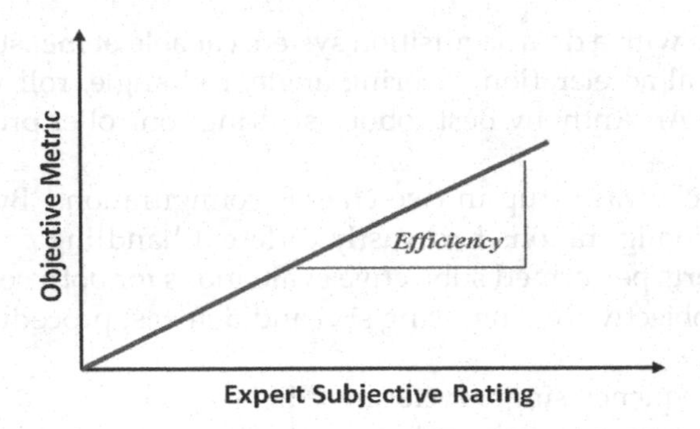


*Figure 9.7* Ideal Function.

system. Readers familiar with DFSS will note this study does not contain noise factors.

Figure 9.7 shows the "ideal function" for optimization. The ideal function represents the desired input to output relationship of the system for optimization. The slope of this function is efficiency. Higher efficiency indicates a more robust test.

## *Control factors*

Control factors are the steering input waveform and vehicle test speed. Three candidate steering waveforms remained after the initial screening study:

- Fixed frequency single cycle sine, A.
- Fixed frequency single cycle sine, B.
- High amplitude triangle pulse with dwell, C.
- Vehicle test speed settings were 45, 60, and 100 mph.

## *Input signal*

The input signal is vehicle chassis performance as subjectively rated by an internal expert jury for the traits of agility, stability, precision, and roll support. Modifications of tires, stabilizer bars, dampers, suspension bushings, and weight distribution provided wide variation in performance (i.e. input signal). The final experiment had nine chassis configurations "input signal settings" for subjective & objective evaluation.

## *Output signal*

The response signals are the calculated objective metrics from each candidate test procedure. Example metrics are peak axle sideslip value, peak roll velocity, and response time lag.

## Experiment

Once the team defined a set of chassis configurations for evaluation and developed data processing code for metrics, the last step was to collect data. An expert jury subjectively rated all nine configurations over a two-day period in the four categories of: agility, stability, precision, and roll support. Following collection of subjective ratings, candidate test procedures were run on each vehicle configuration. Objective testing results were put into DFSS matrices for each transient trait (Figure 9.8). Each test procedure had 6 objective metrics for consideration.

## Data analysis/optimization

Signal to Noise Ratio (SNR) and Efficiency (Beta) are used to define the performance of each test procedure configuration. SNR is a standard measure to compare the level of variation in output response attributable to changing input signal versus that attributable to uncontrolled factors or error. Efficiency is used to quantify the strength of the relationship between input and output response. SNR and Beta were calculated for each control factor+output metric combination and compared for the four performance traits of agility, stability, precision, and roll support.

Based on the DFSS matrix results, the optimum combinations of test procedure and objective metric were chosen for each transient performance attribute (Figure 9.8).



| Control Factor Configurations | Control Factors | | Input Signal "Subjective Performance - Agility" | | | | | | | | | | |
|---|---|---|---|---|---|---|---|---|---|---|---|---|---|
| | waveform | metric | M1 | M2 | M3 | M4 | M5 | M6 | M7 | M8 | M9 | S/N | Beta |
| 1 | A | A | | | | | | | | | | | |
| 2 | A | B | | | | | | | | | | | |
| 3 | A | C | | | | | | | | | | | |
| 4 | B | A | | | | | | | | | | | |
| 5 | B | B | | | | | | | | | | | |
| 6 | B | C | | | | | | | | | | | |
| 7 | C | A | | | | | | | | | | | |
| 8 | C | B | | | | | | | | | | | |
| 9 | C | C | | | | | | | | | | | |

*Figure 9.8* DFSS matrices templates.

| Transient Handling Characteristic | Test Procedure Detail | | |
|---|---|---|---|
| | Steering Waveform | Test Speed | Objective Metric |
| Agility | Fixed frequency single cycle sine, Hz # 1 | 60 mph | Response of lateral acceleration channel |
| Stability | Fixed frequency single cycle sine, Hz # 1 | 60 mph | Response of rear axle sideslip channel |
| Precision | Fixed frequency single cycle sine, Hz # 1 | 60 mph | Rear axle sidslip + lateral acceleration |
| Roll Support | Fixed frequency single cycle sine, Hz # 2 | 60 mph | Response of roll velocity channel |

***Figure 9.9*** DFSS optimized test procedures/metrics.

The final set of test procedures are referred to as the RDT (Robot Driven Transient) tests.

## *Verification and example uses of the RDT test*

The last section of this paper provides some illustrative examples of how objective data from RDT testing can provide additional insight beyond steady-state testing. Three example scenarios are presented for varying: (1) tire performance, (2) wheel specifications, and (3) shock absorber damping rates.

### Example 1 – tire performance

It should not be surprising tire performance has appreciable influence on transient handling performance, but it is not uncommon to find examples where two tires with similar performance specifications (size, cornering stiffness, etc.) result in different subjective performance levels. This is common during the development submission phase of a new tire for an OEM manufacturer. This example is for two tire submissions with virtually identical component performance specifications. Despite this, subjective evaluation revealed a large difference in performance feel. Figures 9.10 and Figure 9.11 plot data from an RDT test for vehicle lateral acceleration and sideslip with each tire. The configurations were run multiple times for repeatability. In the graph "SWANGLE" is the scaled steering angle input simply for reference.

These results show a clear reduction in sideslip with Tire configuration 1 as well as an increase in peak lateral acceleration response.

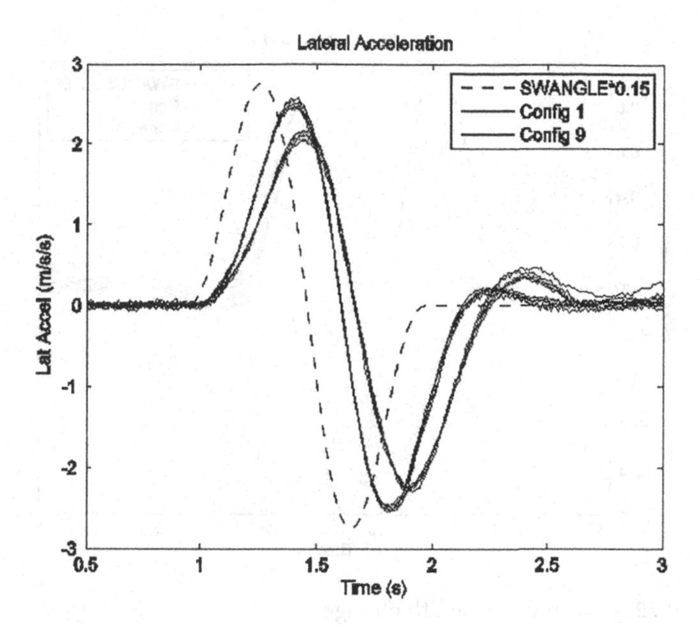


*Figure 9.10* Ay with two tires.

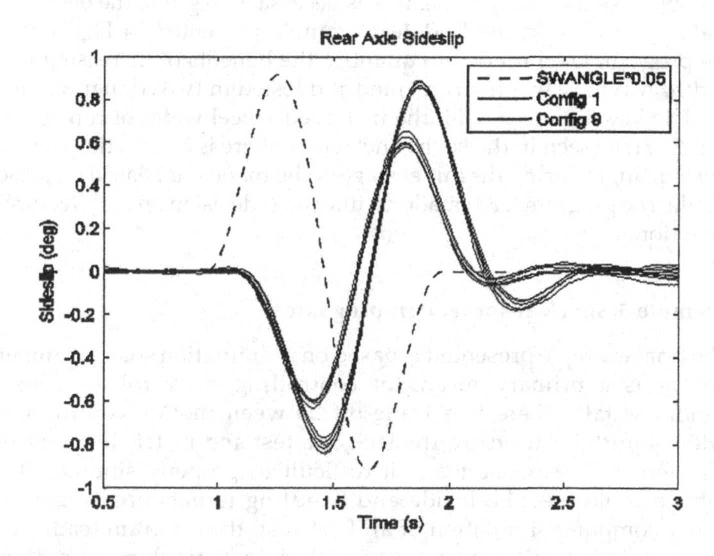


*Figure 9.11* βR with two tires.

**Example 2: wheel specifications**

Wheel width is a common area where there is a distinct trade-off between cost, performance, and mass. Generally, a wider wheel better supports the tire sidewall laterally and improves cornering response,

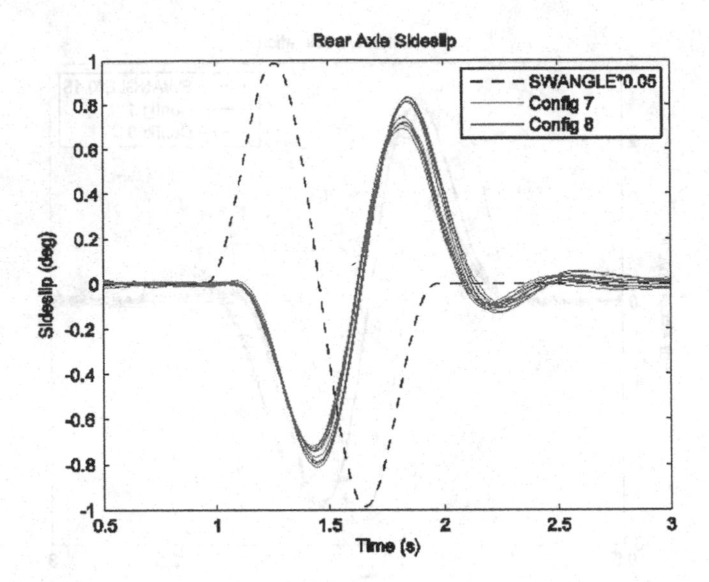


*Figure 9.12* βr with wheel width change.

but objective data supporting this is necessary to guide the decision-making process. In the RDT test example presented in Figure 9.12 the program team needed to quantify the benefits of increasing rim width. A vehicle was instrumented and tested in two configurations.

The key difference with the increased wheel width of configuration B can be seen in the βr channel where there is a reduction in rear sideslip angle during the initial stage of the maneuver. Based on these results the program team made an informed decision on architectural direction.

**Example 3: shock absorber damping rates**

The final example presented is based on a simulation study. Damper tuning is a primary means of controlling body roll response. Fundamentally there is a trade-off between motion control and ride comfort. Historically, the lack of a test and metric to quantify transient roll response made it difficult to properly simulate if a vehicle could meet both ride and handling targets pre-hardware. Using computer simulation of an RDT test, the program team was able to predict roll performance with a realistic damper performance curve. A proprietary computer simulation used this same damping curve for simulation of ride performance. Based on these simulations the damping characteristic was iterated to find a tuning that confirmed it was feasible to meet both ride and handling performance targets.

## Summary

This paper presented objective and subjective evaluation methods to quantify transient vehicle dynamics performance. Subjective evaluations of transient performance were described and categorized into four traits: agility, stability, precision, and roll control. Objective metrics of transient performance were developed using DFSS methods to optimize an open-loop steering robot-controlled test. This newly developed test was used to show examples of how chassis parameters can affect transient performance.

## References

1. Badiru, Ibrahim A. and M. W. Neal (2013), "Use of DFSS Principles to Develop an Objective Method to Assess Transient Vehicle Dynamics," SAE International, Vol. 25, No. 4, pp. 1–8.
2. Milliken, W.F., Milliken, D.L., "Race Car Vehicle Dynamics," SAE International, Warrendale, PA, ISBN 1–56091-526–9, 1995.
3. De Claire, A. "Variable Response Vehicle." GM Research Internal Report, 1968.
4. Crolla, D., Chen, D., Whitehead, J., and Alstead, C., "Vehicle Handling Assessment Using a Combined Subjective-Objective Approach," SAE Technical Paper 980226, 1998, doi:10.4271/980226.
5. Topping, R., "Understeer Concepts with Extensions to Four-Wheel Steer, Active Steer, and Time Transients," SAE Int. J. Passeng. Cars - Mech. Syst. 5(1):167–186, 2012, doi:10.4271/2012–01–0245.
6. J. Lu, D. Messih and A. Salib, "Roll rate based stability control - the Roll Stability Control system," ESV-07–136, Proc. of the 20th Enhanced Safety of Vehicles Conference, 2007.
7. Kodaira, T., Ooki, M., Sakai, H., Katsuyama, E. et al., "Vehicle Transient Response Based on Human Sensitivity," SAE Technical Paper 2008-01-0597, 2008, doi:10.4271/2008–01–0597.
8. Muragishi, Y., Fukui, K., Ono, E., Kodaira, T. et al., "Improvement of Vehicle Dynamics Based on Human Sensitivity (First Report) - Development of Human Sensitivity Evaluation System -," SAE Technical Paper 2007-01-0448, 2007, doi:10.4271/2007–01–0448.

# chapter ten

# Systems modeling for product design

## Introduction

Product design has normally been performed by teams, each with expertise in a specific discipline such as material, structural, and electrical systems. Traditionally, each team would use its member's experience and knowledge to develop the design sequentially. Collaborative design decisions explore the use of optimization methods to solve the design problem incorporating a number of disciplines simultaneously. It is known that the optimum of the product design is superior to the design found by optimizing each disciplines sequentially due to the fact that it is enabled to exploit the interactions between the disciplines. In this paper, a bi-level decentralized framework based on Memetic Algorithm (MA) is proposed for collaborative design decision using forearm crutch as the case. Two major decisions are considered: the weight and the strength. In this chapter, we introduce two design agents for each of the decisions (Wu et al., 2011). At the system level, one additional agent termed facilitator agent is created. Its main function is to locate the optimal solution for the system objective function which is derived from the Pareto concepts; thus, Pareto optimum for both weight and strength is obtained. It is demonstrated that proposed model can converge to Pareto solutions.

Under collaborative design paradigm (Wu et al., 2011), the first common topic is Multidisciplinary Design Optimization (MDO) which is defined as "an area of research concerned with developing systematic approaches to the design of complex engineering artifacts and systems governed by interacting physical phenomena" (Alexandrov, 2005). Researchers agree that interdisciplinary coupling in the engineering systems presents challenges in formulating and solving the MDO problems. The interaction between design analysis and optimization modules and multitudes of users is complicated by departmental and organizational divisions. According to Braun and Kroo (1997), there are numerous design problems where the product is so complex that a coupled analysis driven by a single design optimizer is not practical as the method becomes too time consuming either because of the lead time needed to integrate the analysis or because of the lag introduced by disciplinary sequencing.

Some researchers have taken project management as a means to facilitate and coordinate the design among multi-discipline product design (Thal, Badiru, and Sawhney, 2007).

Early advances in MDO involve problem formulations that circumvent the organizational challenges, one of which is to protect disciplinary privacy by not sharing full information among the disciplines. It is assumed that a single analyst has complete knowledge of all the disciplines. As indicated by Sobeiski and Haftka (1997), most of the work at this phase aims to tackle the problems by a single group of designers within one single enterprise environment where the group of designers shares a common goal and requires less disciplinary optimum. The next phase of MDO gives birth to two major techniques: Optimization by Linear Decomposition (OLD) and Collaborative Optimization (CO). These techniques involve decomposition along disciplinary lines and global sensitivity methods that undertake overall system optimization with minimal changes to disciplinary design and analysis. However, Alexandrov and Lewis (2000) explore the analytical and computational properties of these techniques and conclude that disciplinary autonomy often causes computational and analytical difficulties which result in severe *convergence* problems.

Parallel to these MDO developments, there also evolves the field of Decision-based Design (Huang, 2004; Li et al., 2004; Simpson et al., 2001; Hernandez and Seepersad, 2002; Choi et al., 2003; Wassenaar and Chen, 2003), which provides a means to model the decisions encountered in design and aims at finding "satisfying" solutions (Wassenaar et al., 2005; Nikolaidis, 2007). Research in Decision-Based Design includes the use of adaptive programming in design optimization (Simon, 1955) and the use of discrete choice analysis for demand modeling (Hernandez and Seepersad, 2002; Choi et al., 2003). In addition, there has been extensive research ranging from single-objective Decision-Based Design (Hernandez and Seepersad, 2002) to multi-objective models (Lewis and Mistree, 1998, 1999). It combines Game Theory, Utility Theory, and Decision Sciences for collaborative design, which can be conducted among a group of designers from different enterprises. This technique has several computational difficulties in calculating the "best reply correspondence," and the rational reaction sets especially when the designs are very complex. Besides, several approximations like using response surfaces within these techniques make them prone to errors (Fernandez et al., 2005).

Note that most methods reviewed above have strict assumption on the utility functions and/or constraints (e.g., convexity and quasi-linear of the functions), which limits the application to product design. In this research, we explore the use of a heuristic method, Memetic Algorithm, a combination of Local Search (LS) and Genetic Algorithm (GA) to forearm crutch design which has non-convex objective for one of the decisions.

Forearm crutches had been exclusively used by people with permanent disability. Nowadays, it is beginning to serve for some shorter-term purposes as well. The design of forearm crutch needs to consider multidisciplinary decisions. For example, the structure designer wants to ensure the design is lightweight. The material engineer wants composite material to have the right number of layers at right angles to make the product durable. The outsourcing engineer wants the supplier to provide low cost, high reliable, and lightweight parts. Another important factor impacting the design is cost. Here, we introduce the design agent for each disciplinary decision problem and one system agent facilitating the communication among the design agents and guiding the design to convergence. To achieve this, the overall decision space is partitioned into two sets: one for coupled variables (the ones shared by at least two designers) and the other for local variables (the ones that can be fully controlled by each designer). Next, an iterative process between design agent decisions on local variables and facilitator agent decisions on the whole design space launches. It is demonstrated that a converged Pareto optimum is achieved after a number of iterations for the forearm crutch design which has nonlinear form decision functions.

## Literature review

Collaborative Optimization, introduced by Braun and Kroo (1997), is a bilevel optimization approach where a complex problem is hierarchically decomposed along disciplinary boundaries into a number of subproblems that are brought into multidisciplinary agreement by a system-level coordination process. With the use of local subspace optimizers, each discipline is given complete control over its local design variables subject to its own disciplinary constraints. The system-level problem sets up target values for variables from each discipline. Each discipline sets the objectives to minimize the discrepancy between the disciplinary variable values and the target values. The system-level optimization problem is formulated as minimizing a global objective subject to interdisciplinary consistency constraints. The interdisciplinary consistency constraints are equality constraints that match the system-level variables with the disciplinary variables. In OLD (Sobiesczanaki-Sobieski, 1982, 1988; Sobieszczanski-Sobieski, James, and Dovi, 1985), the disciplines are given the autonomous task of minimizing disciplinary design infeasibility while maintaining system-level consistency. The system-level problem is to drive design infeasibility to zero. At the local-level problem, the disciplines use their local degrees of freedom to minimize the violation of the disciplinary design constraints, subject to matching the target value for the disciplinary output that is fed into the discipline. Balling and Sobieszczanski-Sobieski (1994) introduce a combination of CO and OLD where the disciplinary subproblems minimize the discrepancy in the system-level targets as well as the disciplinary design infeasibility given the disciplinary design constraints.

Both CO and OLD depend on a design problem's amenability to hierarchical decomposition with the system objective explicitly defined. On the other hand, Concurrent Sub-Space Optimization (CSSO) (Sobiesczanaki-Sobieski, 1988) is a nonhierarchic system optimization algorithm that optimizes decomposed subspaces concurrently, followed by a coordination procedure for directing system problem convergence and resolving subspace conflicts. In CSSO, each subspace optimization is a system-level problem with respect to the subset of the total system design variables. Within the subspace optimization, the nonlocal variables that are required to evaluate the objective and the constraint functions are approximated using Global Sensitivity Equations (GSE). Interested readers please refer to Sobiesczanaki-Sobieski (1988), for detailed description of GSEs.

The Bi-Level Integrated Synthesis (BLISS) (Sobiesczanaki-Sobieski and Kodiyalam, 1998) method uses a gradient-guided path to reach the improved system design, alternating between the set of modular design spaces (the disciplinary problems) and the system-level design space. BLISS is an all-in-one like method in that the complete system analysis performed to maintain multidisciplinary feasibility at the beginning of each cycle of the path. The overall problem is decomposed such that a set of local optimization problems deal with the detailed local variables which are large in number and one system-level optimization problem that deals with a small number of global variables.

Decision-Based Design (Fernandez et al., 2002; Hernandez and Seepersad, 2002; Simpson et al., 2001; Choi et al., 2003) is a paradigm focusing on distributed and collaborative design efforts. For the cases where continuous variables are used, adaptive linear programming (Lewis and Mistree, 1999) is employed; in case of mixed discrete and continuous variables, Foraging-Directed Adaptive Linear Programming has been used (Lewis and Mistree, 1999). In a noncooperative environment, game theoretic principles are used to arrive at the best overall design (Lewis and Mistree, 1998, 1999). Recently, Design-for-Market systems grows out of the Decision-Based Design and emerges as an area focusing on establishing a solid basis in decision theory, by taking microeconomics into account, to support engineering design. Kumar et al. (2007) propose a hierarchical choice model based on discrete choice analysis to manage and analyze customer preference data in setting design targets. Azarm's group studies new product designs that are robust from two perspectives – from the engineering perspective in terms of accounting for uncertain parameters and from the market perspective in terms of accounting for variability in customer preferences measurement (Besharati et al., 2006). They conclude incorporating consumer heterogeneity in considering the variability in customer preferences may have significant impact on the ultimate design. Research led by Michalek explores the use of game-theoretic approach to find out market equilibrium under various regulation scenarios (Shiau and Michalek, 2007). A metric for agility measurement is introduced by Sieger, Badiru, and Milatovic (2000) to explore the product development for mass customization.

In general, some common criticisms and/or challenges facing collaborative design decisions are the convergence and information sharing issues:

- Will the decision model converge? If yes, under what condition (assumptions on the function form and design spaces) will it converge? How fast will it converge?
- Most models (CO, OLD, BLISS, etc.) take a top-down approach with the full knowledge of the design space (e.g., the form of utility functions, constraints) being available. For the cases when the design information is partially known, what decision model is appropriate?

To address these challenges, we propose a general decision framework based on MA that allows distributed design teams to arrive at Pareto solutions which is detailed explained in the next section.

## MA and its application to collaborative design

MA is one of the emerging areas in evolutionary computation. It integrates GA with LS to improve the efficiency of searching complex spaces. In MA, GA is used for global exploration, while LS is employed for local exploitation. The complementary nature of GA and LS makes MA an attractive approach for large scale, complex problems, for example, collaborative design.

### Proposed framework for collaborative design

Let us consider a general collaborative design with $z$ design teams. The problem can be represented as

$$\begin{aligned}
&\text{Min } \mathbf{J}(\mathbf{x}, \mathbf{y}) \\
&\text{St. } \mathbf{g}(\mathbf{x}) \le 0 \\
&\quad \mathbf{h}(\mathbf{x}, \mathbf{y}) \le 0 \\
&x_i^{LB} \le x_i \le x_i^{UB} \ (i = 1,\ldots,n_1) \\
&y_i^{LB} \le y_i \le y_i^{UB} (i = 1,\ldots,n_2)
\end{aligned}$$

where $\mathbf{J} = [J_1(\mathbf{x}, \mathbf{y}) \ldots J_Z(\mathbf{x}, \mathbf{y})]^T$, $\mathbf{x} = [x_1 \ldots x_{n1}]^T$, $\mathbf{y} = [y_1 \ldots y_{n2}]^T$, $\mathbf{g} = [g_1(\mathbf{x}) \ldots g_{m1}(\mathbf{x})]^T$, $\mathbf{h} = [h_1(\mathbf{x}, \mathbf{y}) \ldots h_{m2}(\mathbf{x}, \mathbf{y})]^T$, $\mathbf{x}$ is the set of $n_1$ local variables, $\mathbf{y}$ is the set of

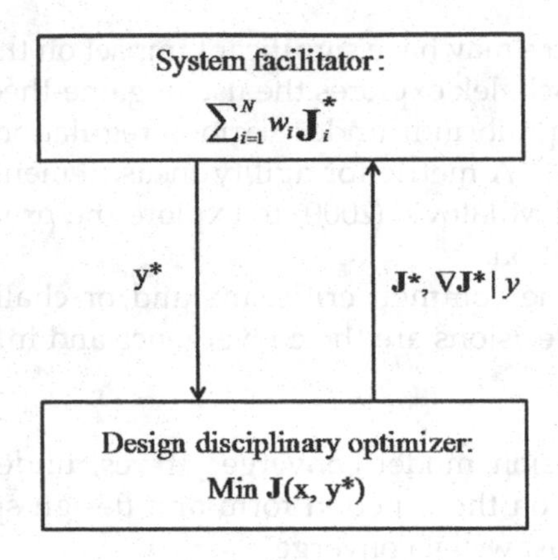


*Figure 10.1* Overall decision framework.

$n_2$ coupled variables, **g** is the set of $m_1$ local constraints, and **h** is the set of $m_2$ coupled constraints.

Figure 10.1 illustrates the iterative decision process between system facilitator agent and disciplinary design agents. First, the facilitator initializes the global solution space over both local and coupled variables. For any solution, e.g., [**x***, **y***], each design agent will execute local optimizer over the sub-design space which consists of **x** only, that is Min **J** (**x**, **y***). The results fed back to the facilitator are the value of objective function and the gradient of objective function over coupled variables. The facilitator will (1) employ local search for the recent results updated by each designer using the related gradient information for the improved design (2) next, traditional GA operators, crossover and mutation, are applied to introduce new candidates to the solution space.

## *Pseudo Code*

This section shows the layout of the Pseudo Code for the proposed methodology as illustrated in Figure 10.2.

Parameters:

$N$: no. of disciplinary design agents;
$w_i$: weight for the objective function of $i$th disciplinary design agent,

where $i = 1, \ldots N$;

$1/W$: weight step size;
$P$: population size;

Pseudo Code Lines

```
(1)  //Initialization
(2)  The set of final Pareto solutions FP = Ø;
(3)  The set of GA population PS = Ø;
(4)  The set of weights combination WS = Ø;
(5)  Given N objective functions, we have Σ_{i=1}^{N} w_i J_i(x_i, y)
(6)  Begin (at facilitator agent level)
(7)     //Enumerate weights combination
(8)     Set w_1 = w_2 = ··· = w_{N-1} = 0;
(9)     Given weight step size 1/W;
(10)      Let each w_i (i = 1,...,N − 1) increases 1/W, w_N = 1 − w_1 − ··· w_{N-1},
            and add (w_1, w_2,..., w_N) to WS;
(11)      //Weights Loop
(12)      For each weights combination (w_1, w_2,..., w_N) in WS, Σ_{i=1}^{N} w_i J_i(x_i, y) is
           constructed;
(13)          //GA loop
(14)          //Initialization
(15)          Generate random population of P solutions and add them to PS;
(16)          For n = 1 to maximum # of generations for GA loop;
(17)              //Crossover and Mutation
(18)              Random select two parents p_a and p_b from PS;
(19)              Generate two offspring p'_a and p'_b by crossover operator;
(20)                 if p'_a and/or p'_b are not feasible, generate new feasible offspring
(21)                 p''_a and/or p''_b using mutation operator;
(22)              //Selection
(23)              Using fitness function (Σ_{i=1}^{N} w_i J_i(x_i, y)) to evaluate the solution, update PS
                   with improved solutions;
(24)              //Local Search Loop
(25)              For each chromosome p_j in PS;
(26)                 Call each Design Agent for local optimization on x (note different
                      optimization engines can be employed based on the design
                      disciplines);
(27)                 Given updates from Design Agent on x, Facilitator agent employs
                      sub-gradient algorithm [19] as local search algorithm to
                      iteratively locate improved solution p'_j with respect to y;
(28)              Next p_j;
(29)              //Pareto Filter:
(30)              For each chromosome p_j in the set PS;
(31)                 If p_j is not dominated by all the solutions in the set FP;
(32)                     Add p_j to the set FP;
(33)                 Else If there are solutions in the set FP are dominated by p_j;
(34)                     ← Remove those solutions in the set FP;
(35)                 End If;
(36)              Next p_j;
(37)          Next n;
(38)  End;
```

*Figure 10.2* Pseudo Code 1.

As shown in the above Pseudo Code, there exist three loops, from outer to inner in the proposed method: weight enumeration (lines 11–37), GA loop (lines 13–37), and local search loop (lines 24–28). That is, given a weights combination (e.g., $w_1 = 0.3$, $w_2 = 0.7$ for two agents), GA is triggered, which applies crossover and mutation operators and selection mechanism (in this case study, elitism selection is employed) for the population update. In addition, for the updated population, local search is further employed to identify improved solutions within the neighborhood. This is achieved by having sub-gradient information from each designer on the coupled variables fed back to the facilitator. Specifically, given any chromosome from the population, each design agent assumes the coupled variables are set and thus conducts optimization on the local variables only. Each design agents would also study the gradients on the coupled variables. Thus, given the values of the coupled variables, both the optimal design on local variables and the sub-gradient on the coupled variables are sent back to the facilitator. Since the priorities of the objective functions reflected by the weight assignments are enumerated exhaustively, all the possible Pareto solutions are located forming the Pareto frontier. In some cases where the priority is known, the weight loop can be removed. Please note that the Pareto filter operation (lines 29–36) is triggered by the facilitator within each weight combination. That is, it is possible that some Pareto solutions given a specific weight may be dominated by the Pareto solutions obtained with other weights.

One distinguishable feature of this proposed approach from other existing methods is that the information is exchanged iteratively between the facilitator and the design agent. For example, passing from the facilitator to the design agent (top-down) is the values of the coupled variable; passing from the design agent back to the facilitator (bottom-up) is the values of the objective function and associated gradient values, passing from the facilitator to the design agents (top-down) is the values of the coupled variables. The main advantage of this approach is a "black box" disciplinary optimizer can be easily plugged in. Second, since the facilitator explores the solution space based on the knowledge of the solution candidates ($\mathbf{x}^*$, $\mathbf{y}^*$), the candidate performance ($\mathbf{J}^*$) instead of the function formulation, a truly decentralized decision without the full knowledge of the design space can be implemented. An industry case is explored to demonstrate the applicability of the proposed framework.

## *Forearm crutch design*

Crutches are mobility aids used to counter mobility impairment or an injury that limits walking ability. Forearm crutches are used by slipping the arm into a cuff and holding the grip (Figure 10.3). It has been increasingly used for patients with shorter-term needs. Earlier study conducted by National Medical Expenditure Survey (NMHS) in 1987 indicates that an estimated 9.5 million (4%) noninstitutionalized U.S. civilians experience difficulty in performing basic life activities; some need crutches for leg support for walking. This number increases due to the baby boomer effect.

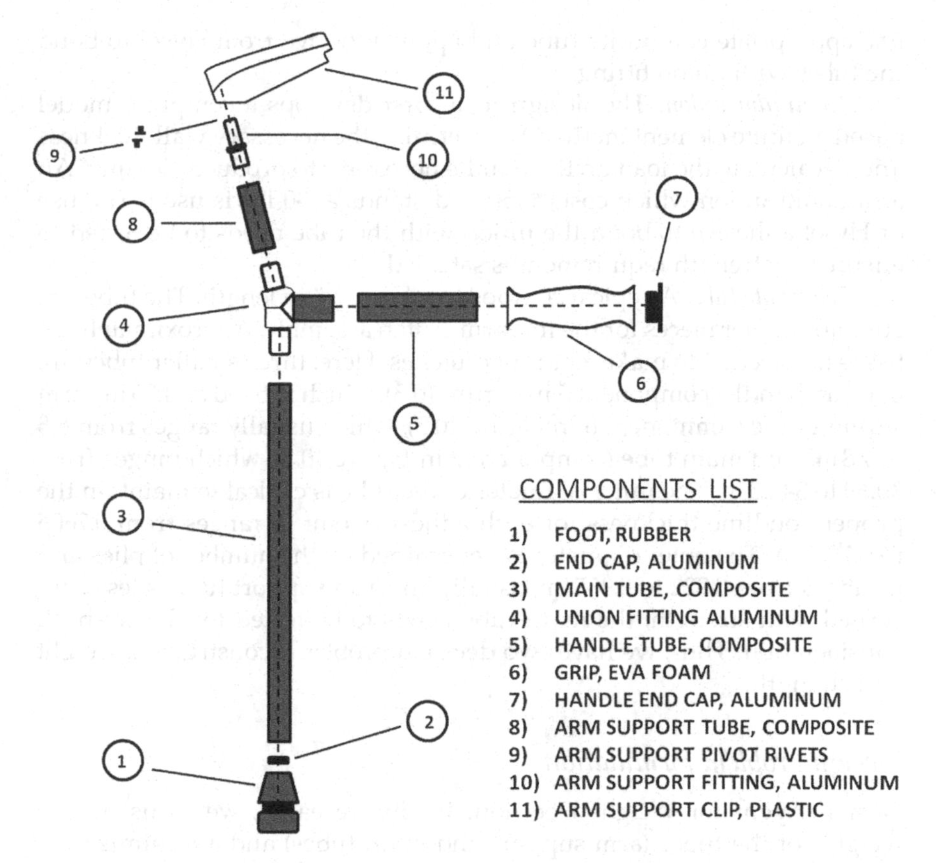


***Figure 10.3*** Exploded view of a forearm crutch.

Typical forearm crutches are made of aluminum and are criticized by customers for being heavy, noisy, and less durable. Customers suggest that a small reduction in the weight of forearm crutches would significantly reduce the fatigue experienced by crutch users. However, the reduction in weight should not be accompanied by a strength reduction. Most crutches on the market are designed for temporary use and wear out quickly. Crutch users commonly have to replace their crutches two to three times a year. This drives the need to redesign forearm crutches which are robust, appropriate for a wide range of users from lighter weight adults to users weighing up to 250 pounds with considerable upper body strength and who may use them aggressively on a continuous basis.

One solution is to use composite material for crutch which is *lightweight* with good performance in *strength*. However, it comes with relatively expensive cost. After in-depth marketing survey, the design team decides to outsource the aluminum union fitting (component #4 in Figure 10.3), use appropriate composite tube and apply adhesive from Hysol to bond the tubes with union fitting.

*Aluminum union:* The design team first develops a computer model based on finite element method to determine the necessary wall thickness and to calculate the load on the handle necessary to produce yielding. An aluminum union which costs $150 and stands ≥630 lbs is used. The use of Hysol adhesive to bond the union with the tube needs to be tested to ensure the strength requirement is satisfied.

*Composite tube:* A typical composite tube is 39" in length. The tube can cut into smaller pieces for the forearm crutch assembly. Approximately 2½ tubes are needed to make a pair of crutches. Here, three smaller tubes are used as: handle (component 5 in Figure 10.3), which is fixed as 4.75 in.; arm support tube (component 8 in Figure 10.3), which usually ranges from 6.5 to 7.8 in.; and main tube (component 3 in Figure 10.3), which ranges from 30.69 to 34.25 in. The inner diameter of the tube is critical to maintain the proper bondline thickness for each adhesive joint. It ranges from 0.7605 to 0.7735 in. The outer diameter is determined by the number of plies and it ranges from 0.922 to 0.935 in. Usually, the arm support tube is less concerned with strength, the main tube needs to be tested for the strength consideration. Thus, we have two decision problems constructed: weight and strength.

## *Design problem formulation*

Design Agent for Weight Decision: In this research, we focus on the weights of the tubes (arm support and main tubes) and a minimization problem is introduced as

$$\text{Min}: W = W_u + W_L$$

$$\text{St}: W_u = \rho\pi\left[\left(\frac{D_o}{2}\right)^2 - \left(\frac{D_i}{2}\right)^2\right] \times L_u$$

$$W_L = \rho\pi\left[\left(\frac{D_o}{2}\right)^2 - \left(\frac{D_i}{2}\right)^2\right] \times L_L$$

$$30.6875 \le L_L \le 34.25$$

$$6.5 \le L_u \le 7.8$$

$$0.922 \le D_o \le 0.935$$

$$0.7605 \le D_i + 2T/1{,}000 \le 0.7735$$

where $W_u$ (in.) is the weight of arm support tube; $W_L$ (in.) is the weight of main tube; $\rho$ (lbs/in.$^3$) is the density of the composite tube, which is 0.08 in this paper; $L_u$ (in.) is the length of the arm support tube; $L_L$ (in.) is the length of the main tube; $D_o$ (in.) is the outer diameter; $D_i$ (in.) is the inner diameter; and $T$ (mils, 1 mils = 0.001 in.) is the bondline adhesive material thickness.

### *Design agent for strength decision*

Since the strength from aluminum fitting is satisfactory from Finite Element Analysis (FEM), the strength model will consider two potential failures: the adhesive applied joint and the strength of the main tube. Thus, the problem is constructed as

$$\text{Max}: S = \text{Min}(S_L, S_A)$$

$$\text{St}:$$

$$S_L = \frac{\pi EI}{L_L^2}$$

$$I = \pi\left(D_o^4 - D_i^4\right)/64$$

$$12 \le E \le 16$$

$$S_A = \left(-6.0386T^2 + 7.7811T + 4644.5\right) \times \frac{\pi}{4} \times \left(D_o^2 - D_i^2\right)$$

$$0.922 \le D_o \le 0.935$$

$$0.7605 \le D_i + 2T/1{,}000 \le 0.7735$$

$$30.6875 \le L_L \le 34.25$$

$$0 \le T \le 17$$

where $S_L$ (lbs) is the strength of the bottom of the lower tube, $E$ (msi, 1 msi = $10^6$ psi) is the modulus of elasticity, $I$ (in.$^4$) is the area moment of inertia, and $S_A$ (lbs) is the strength of the joint after applying adhesive.

## *Implementation*

For the decision problems explained above, optimization code written in MATLAB® is executed. Here, we provide detailed explanation how the system problem is constructed and how the facilitator agent guides the design agents to converge to the solution using MA.

*Step 1 Initialization:* Given $w_1$, $w_2$, construct system search space as $w_1W^*-w_2S^*$ ($W^*$ and $S^*$ are the values of the objectives from each design agent, $w_1 + w_2 = 1$).

*Step 2 Real Code Genetic Algorithm:* The chromosome is represented with real numbers, that is ($L_u$, $L_L$, $D_o$, $D_i$, $T$, $E$). Note: $L_L$, $D_o$, $D_i$, $T$ are coupled variables, $L_u$ is the local variable for weight agent, and $E$ is the local variable for strength agent.

*Step 2.1 (Initial population):* For ($L_u$, $L_L$, $D_o$, $D_i$, $T$, $E$), without losing of generalization, let assume $a$ and $b$ represent the lower bound and upper bound of one of the variable, $r$ be a random number $r \in [0,1]$, we get $(b - a)\ r + a$. Thus, a new chromosome is generated as for the initial population. A pool of 40 chromosomes is created.

*Step 2.2 (Selection of parents):* To ensure all chromosomes have the chances to be selected, solutions are classified into three groups according to their fitness: high fitness level, medium fitness level, and low fitness level. The fitness is assessed based on $w_1W^*-w_2S^*$, the lower, the better.

*Step 2.3 (Crossover):* Given two chromosome $C_1 = \left(L_u^1, L_L^1, D_o^1, D_i^1, T^1, E^1\right)$ and $C_2 = \left(L_u^2, L_L^2, D_o^2, D_i^2, T^2, E^2\right)$, the offspring are generated as:

$$C_1' = \theta C_1 + (1-\theta) C_2$$

$$C_2' = (1-\theta) C_1 + \theta C_2$$

where $\theta[0,1]$

*Step 2.4 (Mutation):* Mutation is applied by simply generating a new feasible solution to replace the infeasible one.

*Step 3 (Local Search):* The facilitator agent applies sub-gradient method based LS over coupled variables to improve the solutions. First, each design agent evaluates the gradients of the design decision problems (disciplinary) *wrt* the coupled variables. For example, given the coupled variables $L_L = L_L^*, D_o = D_o^*, D_i = D_i^*, T = T^*$, each decision problem is solved independently for $W^*$ and $S^*$. The gradients are obtained as:

$$\lambda_{w,L_L} = \frac{\partial W}{\partial L_L}\big|_{L_L = L_L^*, D_o = D_o^*, D_i = D_i^*, T = T^*}, \quad \lambda_{S,L_L} = \frac{\partial S}{\partial L_L}\big|_{L_L = L_L^*, D_o = D_o^*, D_i = D_i^*, T = T^*}$$

$$\lambda_{w,D_o} = \frac{\partial W}{\partial D_o}\big|_{L_L = L_L^*, D_o = D_o^*, D_i = D_i^*, T = T^*}, \quad \lambda_{S,D_o} = \frac{\partial S}{\partial D_o}\big|_{L_L = L_L^*, D_o = D_o^*, D_i = D_i^*, T = T^*}$$

$$\lambda_{w,D_i} = \frac{\partial W}{\partial D_i}\Big|_{L_L=L_L^*,D_o=D_o^*,D_i=D_i^*,T=T^*},\ \lambda_{S,D_i} = \frac{\partial S}{\partial D_i}\Big|_{L_L=L_L^*,D_o=D_o^*,D_i=D_i^*,T=T^*}$$

$$\lambda_{w,T} = \frac{\partial W}{\partial T}\Big|_{L_L=L_L^*,D_o=D_o^*,D_i=D_i^*,T=T^*},\ \lambda_{S,T} = \frac{\partial S}{\partial T}\Big|_{L_L=L_L^*,D_o=D_o^*,D_i=D_i^*,T=T^*}$$

The gradients of the system problem are then calculated as

$$\lambda_{L_L} = w_1\lambda_{w,L_L} - w_2\lambda_{S,L_L}$$

$$\lambda_{D_o} = w_1\lambda_{w,D_o} - w_2\lambda_{S,D_o}$$

$$\lambda_{D_i} = w_1\lambda_{w,D_i} - w_2\lambda_{S,D_i}$$

$$\lambda_T = w_1\lambda_{w,T} - w_2\lambda_{S,T}$$

Based on $\lambda = [\lambda_{L_L}, \lambda_{D_o}, \lambda_{D_i}, \lambda_T]$, the facilitator agent non-summable diminishing method to update the coupled variables. That is, at iteration $k+1$,

$$\begin{bmatrix} L_L \\ D_o \\ D_i \\ T \end{bmatrix}^{(k+1)} = \begin{bmatrix} L_L \\ D_o \\ D_i \\ T \end{bmatrix}^{(k)} - \alpha_{k+1}\begin{bmatrix} \lambda_{L_L} \\ \lambda_{D_o} \\ \lambda_{D_i} \\ \lambda_T \end{bmatrix}^{(k)}$$

where step size $\alpha_k$ satisfies:

$$\begin{cases} \lim_{k\to\infty} \alpha_k = 0 \\ \sum_{k=1}^{\infty} \alpha_k = \infty \end{cases}$$

The coupled variables are updated based on the above sub-gradient method until no further improvement of the weighted system problem is required.

## *Results and analysis*

The Pareto frontier obtained by the proposed decentralized framework is shown in Figure 10.4. Note that the problem has Min–Max structure. Since this project focuses on the composite tube design (main tube and handle tube), the weight for the handle tube (component #5) is computed as

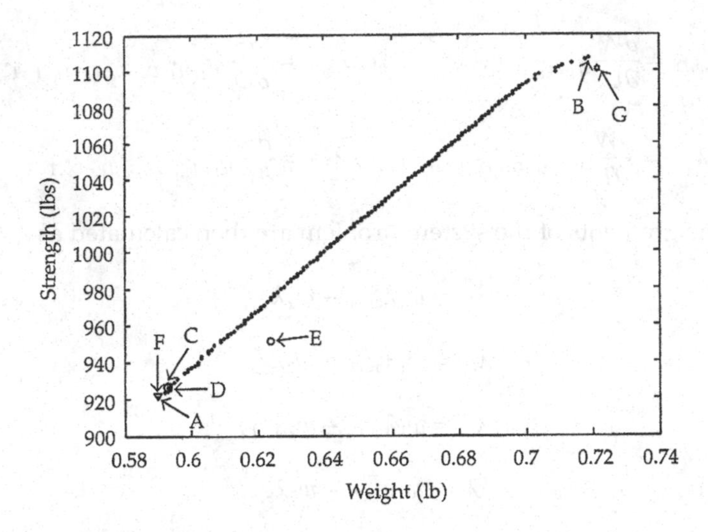


***Figure 10.4*** Pareto frontier in performance space for the crutch design.

$$\rho\pi\left[\left(\frac{D_o}{2}\right)^2-\left(\frac{D_i}{2}\right)^2\right]\times 4.75$$

Other components in Figure 10.3 are outsourced with the weights summarized in Table 10.1.

We choose Pareto solution (A and B) to compare with the composite crutch from Ergonomics and the Invacare crutch which are two commercial products in Table 10.2. Apparently, most composite crutches outperform Ergonomics and Invacare for both weight and strength except Design B outweighs Ergonomics by 0.09 lbs. However, Design B is much durable with strength being 1,105 lbs compared to 715 lbs of Ergonomics.

It is expected that the cost of the composite crutch will be high. In this case, it is around $460 in total (tube and other components shown in Figure 10.3). The crutch produced by Invacare and Ergonomics price range is $60–$250. Although the composite crutch is several times expensive,

***Table 10.1*** Weight for each component of the crutch

| Components (Figure 10.3) | Weight (lbs) |
|---|---|
| #2 | 0.006 |
| #4 | 0.05 |
| #7 | 0.0074 |
| #10 | 0.025 |
| Others (#1, #6, #9, #11) | 0.2 |

***Table 10.2*** Comparison of crutch weight and strength

| Crutch design | Weight (lbs) | Strength (lbs) |
|---|---|---|
| Invacare | 2.3 | 630 |
| Ergonomics | 1 | 715 |
| Pareto design (A) | 0.9498 | 921 |
| Pareto design (B) | 1.0945 | 1,107 |
| Nash equilibrium (C) | 0.9532 | 926 |
| Weight leader strength follower (D) | 0.9532 | 926 |
| Strength leader weight follower (E) | 0.9879 | 951 |
| Weight complete control (F) | 0.9499 | 922 |
| Strength complete control (G) | 1.0952 | 1,100 |

it lasts much longer. Instead of having replacement two to three times a year, it can be used for a number of years since the lighter composite crutch could sustain greater than 1,100 pounds load.

## Conclusion

Collaborative design decisions involve designers from different discipline with different specific domain knowledge. The decision process is a sequence of phases or activities where mathematical modeling can employ. In this chapter, a bi-level distributed framework based on MA is proposed. Since the information communicated is neither the form of the decision function nor the decision space, private information is protected. In addition, in the cases where the information is not complete, the proposed framework can still guarantee the convergence to Pareto solutions. To demonstrate the applicability of the framework, a forearm clutch design is studied in details. The results confirm converged Pareto set can be obtained for any form of decision function. While promising, the decision problem constructed are deterministic, our next step is to explore the use of this framework for design decisions under uncertainty. Computational efficient approach in the area of reliability based design optimization would be explored.

## References

Alexandrov, N.M. (2005). Editorial – multidisciplinary design optimization, *Optimization and Engineering* 6(1), 5–7.

Alexandrov, N.M. and Lewis, R.M. (June 2000). An analysis of some bilevel approaches to multidisciplinary design optimization. Technical report, Institute for Computer Applications in Science and Engineering, Mail Stop 132C, NASA Langley Research Center, Hampton, Virginia 23681-2199.

Balling, R. and Sobieszczanski-Sobieski, J. (September 1994). Optimization of coupled systems: A critical overview. *AIAA-94-4330, AIAA/NASA/USAF/ ISSMO 5th Symposium on Multidisciplinary Analysis and Optimization*, Panama City Beach, Florida, publication in AIAA J., 1995.

Besharati, B., Luo, L., Azarm, S., and Kannan, P.K. (2006). Multi-objective single product robust optimization: An integrated design and marketing approach, *Journal of Mechanical Design* 128(4), 884–892.

Braun, R.D. and Kroo, I.M. (1997). Development and application of the collaborative optimization architecture in a multidisciplinary design environment. Multidisciplinary Design Optimization: State of the Art, SIAM, pp. 98–116.

Choi, H., Panchal, J.H., Allen, K.J., Rosen, W.D., and Mistree, F. (2003). Towards a standardized engineering framework for distributed collaborative product realization. *Proceedings of Design Engineering Technical Conferences and Computers and Information in Engineering Conference*, Chicago, IL, USA.

Fernandez, M.G., Panchal, J.H., Allen, J.K., and Mistree, F. (2005). An interactions protocol for collaborative decision making - concise interactions and effective management of shared design spaces. *ASME Design Engineering Technical Conferences and Computer and Information in Engineering Conference*, Long Beach, CA. Paper No. DETC2005–85381.

Fernandez, G.M., Rosen, W.D., Allen, K.J., and Mistree, F. (4–6 September 2002). A decision support framework for distributed collaborative design and manufacture. *9th AIAA/ISSMO Symposium on Multidisciplinary Analysis and Optimization*, Atlanta, Georgia.

Hernandez, G. and Seepersad, C.C. (2002). Design for maintenance: A game theoretic approach, *Engineering Optimization* 34(6), 561–577.

Huang, C.C. (2004). A multi-agent approach to collaborative design of modular products, *Concurrent Engineering: Research and Applications* 12(2), 39–47.

Kumar, D., Hoyle, C., Chen, W., Wang, N., Gomez-levi, G., and Koppelman, F. (September 2007). Incorporating customer preferences and market trends in vehicle packaging design. *Proceedings of the DETC 2007, ASME 2007 International Design Engineering Technical Conferences and Computers and Information in Engineering Conference*, Las Vegas, NV, USA.

Lewis, K. and Mistree, F. (1998). Collaborative, sequential and isolated decisions in design, *ASME Journal of Mechanical Design* 120(4), 643–652.

Lewis, K. and Mistree, F. (1999). FALP: Foraging directed adaptive linear programming. A hybrid algorithm for discrete/continuous problems, *Engineering Optimization* 32(2), 191–218.

Li, W.D., Ong, S.K., Fuh, J.Y.H., Wong, Y.S., Lu, Y.Q., and Nee, A. (2004). Feature –based design in a distributed and collaborative environment, *Computer-Aided Design* 36, 775–797.

Nikolaidis, E. (2007). Decision-based approach for reliability design, *ASME Journal of Mechanical Design* 129(5), 466–475.

Shiau, C.S. and Michalek, J.J. (2007). A game-theoretic approach to finding market equilibria for automotive design under environmental regulation. *Proceedings of the ASME International Design Engineering Technical Conferences*, Las Vegas, NV, USA.

Sieger, D., Badiru, A., and Milatovic, M. (2000). A metric for agility measurement in product development, *IIE Transactions* 32, 637–645.

Simon, H. (1955). A behavioral model of rational choice, *Quarterly Journal of Economics* 6, 99–118.

Simpson, T.W., Seepersad, C.C., and Mistree, F. (2001). Balancing commonality and performance within the concurrent design of multiple products in a product family, *Concurrent Engineering Research and Applications* 9(3), 177–190.

Sobieszczanski-Sobieski, J. (1982). *A Linear Decomposition Method for Large Optimization Problems Blueprint for Development.* National Aeronautics and Space Administration, NASA/TM-83248-1982, Cleveland, OH.

Sobieszczanski-Sobieski, J. (September 1988). Optimization by decomposition: A step from hierarchic to nonhierarchic systems. Technical Report TM 101494, NASA, Hampton, VA.

Sobieszczanski-Sobieski, J. and Haftka, R.T. (August 1997). Multidisciplinary aerospace design optimization: Survey of recent developments, *Structural Optimization* 14(1), 1–23.

Sobieszczanski-Sobieski, J., James, B., and Dovi, A. (1985). Structural optimization by multilevel decomposition, *AIAA Journal* 23, 1775–1782.

Sobieszczanski-Sobieski, J. and Kodiyalam, S. (1998). BLISS/S: A new method for two-level structural optimization, *Structural Multidisciplinary Optimization* 21, 1–13.

Thal, A.E., Badiru, A., and Sawhney, R. (2007). Distributed project management for new product development, *International Journal of Electronic Business Management* 5(2), 93–104.

Wassenaar, H. and Chen, W. (2003). An approach to decision-based design with discrete choice analysis for demand modeling, *ASME Journal of Mechanical Design* 125(3), 480–497.

Wassenaar, H., Chen, W., Cheng, J., and Sudjianto, A. (2005). Enhancing discrete choice demand modeling for decision-based design, *ASME Journal of Mechanical Design* 127(4), 514–523.

Wu, T., Soni, S., Hu, M., Li, F., and Badiru, A. (2011). The application of memetic algorithms for forearm crutch design: A case study, *Mathematical Problems in Engineering* 3(4), 1–15.

# chapter eleven

# Dynamic fuzzy systems modeling

## Introduction – decision support systems, uncertainties

This chapter, based on Ogunwolu (2005), Ibidapo-Obe and Ogunwolu (2004), and Ibidapo-Obe and Asaolu (2006), discusses scientific uncertainty, fuzziness, and application of Stochastic-Fuzzy models in Urban Transit, Water Resources, Energy Planning, and in Education (Universities Admission Process and other HEIs (Higher Educational Institutes) in developing economies). It enunciates the prime place of Decision Support Systems (DSS) Models in providing a robust platform for enabled action on developmental issues. Scientists now recognize the importance of studying scientific phenomenon having complex interactions among their components. These components include not only electrical or mechanical parts but also "soft science" (human behavior, etc.) and how information is used in models. Most real-world data for studying models are uncertain. Uncertainty exists when facts, state, or outcome of an event cannot be determined with probability of 1 (in a scale of 0–1). If uncertainty is not accounted for in model synthesis and analysis, deductions from such models become at best uncertain. The "lacuna" in understanding the concept of uncertainty and developmental policy formulation/implementation are not only due to the non-acceptability of its existence in policy foci but also the radically different expectations and modes of operation that scientists and policymakers use. It is therefore necessary to understand these differences and provide better methods to incorporate uncertainty into policymaking and developmental strategies (Figure 11.1).

Scientists treat uncertainty as a given, a characteristic of all data and information (as processed data). Over the years, however, sophisticated methods to measure and communicate uncertainty, arising from various causes, have been developed. In general, more uncertainty has been uncovered rather than the absolute precision. Scientific inquiry can only set boundaries on the limits of knowledge. It can define the edges of the envelope of known possibilities, but often the envelope is very large and the probabilities of the content (the known possibilities) occurring can be a complete mystery. For instance, scientists can describe the range of uncertainty about global warming and toxic chemicals and perhaps about the relative probabilities of different outcomes, but in most important

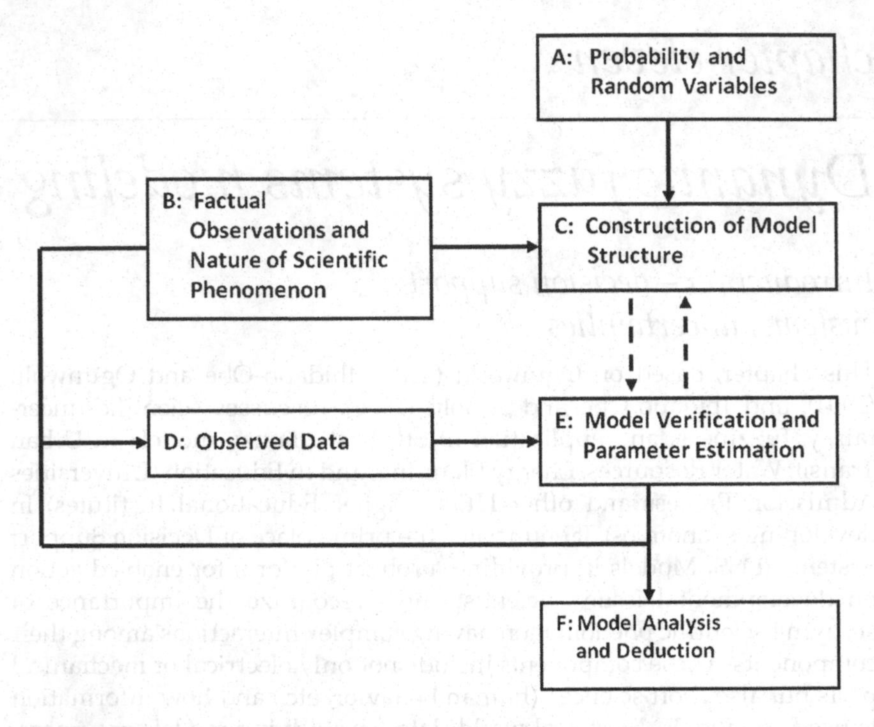


***Figure 11.1*** Basic cycle of probabilistic modeling and analysis.

cases, they cannot say which of the possible outcomes will occur at any particular time with any degree of accuracy. Current approaches to policymaking, however, try to avoid uncertainty and gravitate to the edges of the scientific envelope. The reasons for this bias are clear. Policymakers want to make unambiguous, defensible decisions that are to be codified into laws and regulations.

Although legislative language is often open to interpretation, regulations are much easier to write and enforce if they are stated in absolutely certain terms. Science defines the envelope while the policy process gravitates to an edge – usually the edge that best advances the policymaker's political agenda! But to use science rationally to make policy, the whole envelope and all of its contents must be considered.

## *Decision support systems*

A decision is a judgment, choice, resolution on a subject matter made after due consideration of alternatives or options. It involves setting the basic objectives, optimizing the resources, determining the main line of strategy, planning, and coordinating the means to achieve them, managing relationships and keeping things relevant in the operating environment.

A competitive situation exists when conflicting interests must be resolved. A good decision must be auditable; that is, it must be the best decision at the time it was taken and must have been based on facts or assumptions as well as other technical, sociopolitical, and cultural considerations.

DSS are computer-based tools from problem formulation through solution, simulation implementation, archiving, and reporting. These tools include Operations Research Software Tools, Statistical and Simulation Programs, Expert Systems, and Project Management Utilities. DSS scheme is as follows:

- Problem statement (modeling–dry testing)
- Data information requirements (collection and evaluation procedure)
- Performance measure (alternatives-perceived worth/utility)
- Decision model (logical framework for guiding project decision)
- Decision-making (real world situation, sensitivity analysis)
- Decision implementation (schedule and control)

## *Uncertainty*

In science, information can be, for example, objective, subjective, dubious, incomplete, fragmentary, imprecise, fluctuating, linguistic, data-based, or expert-specified. In each particular case, this information must be analyzed and classified to be eligible for quantification. The choice of an appropriate uncertainty model primarily depends on the characteristics of the available information. In other words, the underlying reality with the sources of the uncertainty dictates the model.

A form of uncertainty stems from the variability of the data. In equal and or comparable situations, each datum in question may not show identical states. This kind of uncertainty is typically found in data taken from plants and animals and reflects the rich variability of nature. Another kind of uncertainty is the impossibility of observing or measuring to a certain level of precision. This kind of precision depends not only on the power of the sensors applied but also on the environment including the observer. This type of uncertainty can also be termed as uncertainty due to partial ignorance or imprecision. Finally, uncertainty is introduced by using a natural or professional language to describe the observation as a datum. This vagueness is a peculiar property of humans and uses the special structure of human thinking. Vagueness becomes more transparent in a case, such as when dealing with grades, shades, or nuances, expressed verbally and represented by marks or some natural numbers. Typical phrases used in such vague descriptions include "in many cases," "frequently," "small," "high," "possibly," "probably," etc. All these kinds of uncertainty (uncertainty due to variability, imprecision, and vagueness) can also occur in combinations.

One approach is to investigate uncertainty by use of Sensitivity Analysis whereby the given data are subjected to small variations to see how these variations will influence the conclusions drawn from the data. The problem with sensitivity analysis, however, has to do with its "point-oriented" approach as to where and in what dimensions the variations are to be fixed. Another approach is in the use of Interval Mathematics, in which each datum is replaced by a set of "possible" surrounding data on the real line. The problem with interval mathematics is the difficulty of specifying sharp boundaries for the data sets, e.g., the ends of the intervals. A third approach for tackling uncertainty is the stochastic approach. This involves realization of each datum as a random variable in time, i.e., the datum is assumed to be chosen from a hypothetical population according to some fixed probability law. This approach works well with modeling of variability and small imprecision. Finally, uncertainty can be taken into account using notions and tools from fuzzy set theory. In this approach, each datum is represented by a fuzzy set over a suitable universe. The main idea of fuzzy set is the allowance of membership, to a grade, for every element of a specified set. With this notion, uncertainty can be modeled mathematically, more adequately and subtly, using only the common notion of membership of an element to a set. Fuzzy set models both imprecision and vagueness.

However, in typical real world systems and decision-making processes, virtually all the three types of uncertainty (variability, imprecision, and vagueness) manifest. Since stochasticity captures variability and small imprecision well and fuzzy set captures imprecision and vagueness in data description well, then for a comprehensive treatment of uncertainty in data, it is advisable to exploit a combined effect of stochastic and fuzzy types of uncertainty. Simply put, Randomness caters for objective data while Fuzziness caters for subjective data.

## *Fuzziness*

Fuzzy Logic (and reasoning) is a scientific methodology for handling uncertainty and imprecision. Unlike in conventional (crisp) sets, the members of fuzzy sets are permitted varying degrees of membership. An element can belong to different fuzzy sets with varying membership grade in each set. The main advantage of fuzzy sets is that it allows classification and gradation to be expressed in a more natural language; this modeling concept is a useful technique when reasoning in uncertain circumstances or with inexact information that is typical of human situations. Fuzzy models are constructed based on expert knowledge rather than on pure mathematical knowledge; therefore, they are both quantitative and qualitative, but are considered to be more qualitative than quantitative. Therefore, a fuzzy expert system is a computer based decision tool that manipulates imprecise inputs based on the knowledge of an expert in that domain.

A Fuzzy Logic Controller (FLC) makes control decisions by its well-known fuzzy IF–THEN rules. In the antecedence of the fuzzy rules (i.e., the IF part), the control space is partitioned into small regions with respect to different input conditions. Membership Function (MF) is used to fuzzify each of the input variables. For continuity of the fuzzy space, the regions are usually overlapped by their neighbors. By manipulating all the input values in the fuzzy rule base, an output will be given in the consequent (i.e., the THEN part). FLCs can be classified into two major categories: the Mamdani (M) type FLC that uses fuzzy numbers to make decisions and the Takagi–Sugeno (TS) type FLC that generates control actions by linear functions of the input variables.

## *The fuzzy set specifications*

In classical set theory, the membership of elements in relation to a set is assessed in binary terms according to a crisp condition. An element either belongs or does not belong to the set; the boundary of the set is crisp. As a further development of classical set theory, fuzzy set theory permits the gradual assessment of the membership of elements in relation to a set; this is described with the aid of an MF.

If X represents a fundamental set and $x$ are the elements of this fundamental set, to be assessed according to an (lexical or informal) uncertain proposition and assigned to a subset $A$ of X, the set

$$\tilde{A} = \{(x, \mu_A(x)) \mid x \in X\}$$

is referred to as the uncertain set or fuzzy set on X (Figure 11.2).

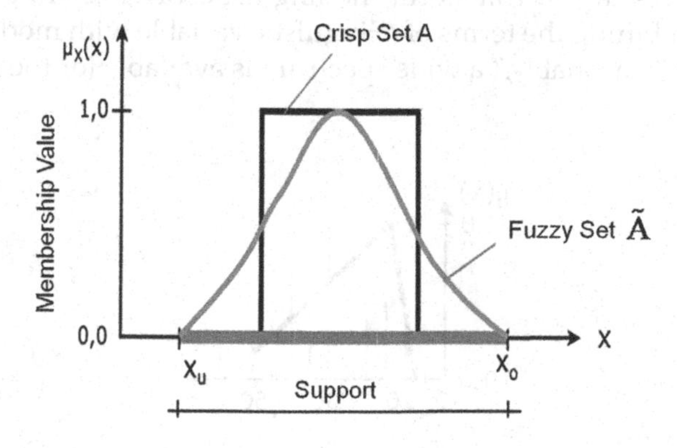


***Figure 11.2*** Fuzzy set.

$\mu_A(x)$ is the membership function of the fuzzy set and may be continuous or discrete with elements assessed by membership values.

The uncertainty model fuzziness lends itself to describing imprecise, subjective, linguistic, and expert-specified information. It is capable of representing dubious, incomplete, and fragmentary information and can additionally incorporate objective, fluctuating, and data-based information in the fuzziness description. Requirements regarding special properties of the information do generally not exist. With respect to the regularity of information within the uncertainty, the uncertainty model fuzziness is less rigorous in comparison with probabilistic models. It specifies lower information content and thus possesses the advantage of requiring less information for adequate uncertainty quantification.

Primarily, fuzzification (Figure 11.3) is a subjective assessment, which depends on the available information. In this context, four types of information are distinguished to formulate guidelines for fuzzification. If the information consists of various types, different fuzzification methods may be combined.

*Information type I: Sample of very small size*

The MF is specified on the basis of existing data comprising elements of a sample. The assessment criterion for the elements $x$ is directly related to numerical values derived from X. An initial draft for an MF may be generated with the aid of simple interpolation algorithms applied to the objective information, e.g., represented by a histogram. This is subsequently adapted, corrected, or modified by means of subjective aspects.

*Information type II: Linguistic assessment*

The assessment criterion for the elements $x$ of $X$ may be expressed using linguistic variables and associated terms, such as "low" or "high" as shown in Figure 11.4. As numerical values are required for a fuzzy analysis, it is necessary to transform the linguistic variables to a numerical scale. By combining the terms of a linguistic variable with modifiers, such as "very" or "reasonably," a wide spectrum is available for the purpose of assessment.

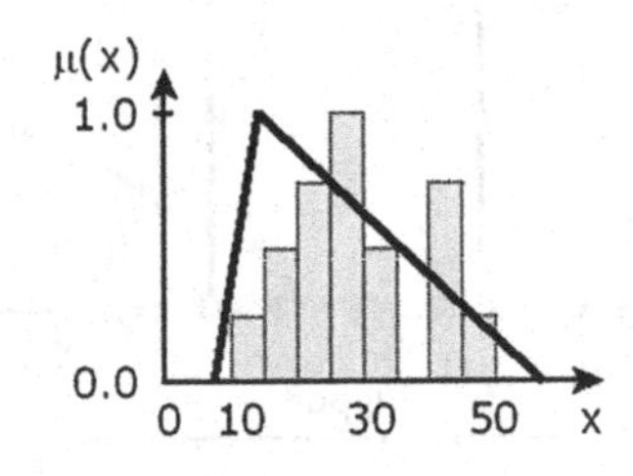


***Figure 11.3*** Fuzzification of information from a very small sample.

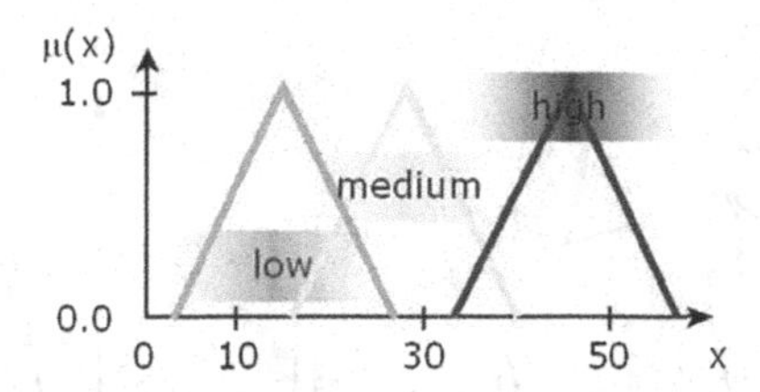


***Figure 11.4*** Fuzzification of information from a linguistic assessment.

*Information type III: Single uncertain measured value*

If only a single numerical value from $X$ is available as an uncertain result of measurement $m$, the assessment criterion for the elements $x$ may be derived from the uncertainty of the measurement, which is quantified on the assigned numerical scale. The uncertainty of the measurement is obtained as a "gray zone" comprising more or less trustworthy values. This can be induced, e.g., by the imprecision of a measurement device or by a not clearly specifiable measuring point.

The experimenter evaluates the uncertain observation for different membership levels. For the level $\mu_A(x) = 1$, a single measurement or a measurement interval is specified in such a way that the observation may be considered to be "as crisp as possible." For the level of the support, $\mu_A(x) = 0$, a measurement interval is determined that contains all possible measurements within the scope of the observation. An assessment of the uncertain measurements for intermediate levels is left up to the experimenter. The MF is generated by interpolation or by connecting the determined points $(x, \mu_A(x))$. Figure 11.5 shows an example.

*Information type IV: Knowledge based on experience*

The specification of an MF generally requires the consideration of opinions of experts or expert groups, of experience gained from comparable problems, and of additional information where necessary. Also, possible errors in measurement, and other inaccuracies attached to the fuzzification process may be accounted for. These subjective aspects generally supplement the initial draft of an MF. If neither reliable data nor linguistic assessments are available, fuzzification depends entirely on estimates by experts.

As an example, consider a single measurement carried out under dubious conditions, which only yields some plausible value range. In those cases, a crisp set may initially be specified as a kernel set of the fuzzy set. The boundary regions of this crisp kernel set are finally "smeared" by assigned membership values $\mu_A(x) < 1$ to elements close to the boundary and leading the branches of $\mu_A(x)$ beyond the boundaries of the crisp kernel set monotonically to $\mu_A(x) = 0$. By this means, elements that do not belong to the crisp kernel set, but are located "in the neighborhood" of the

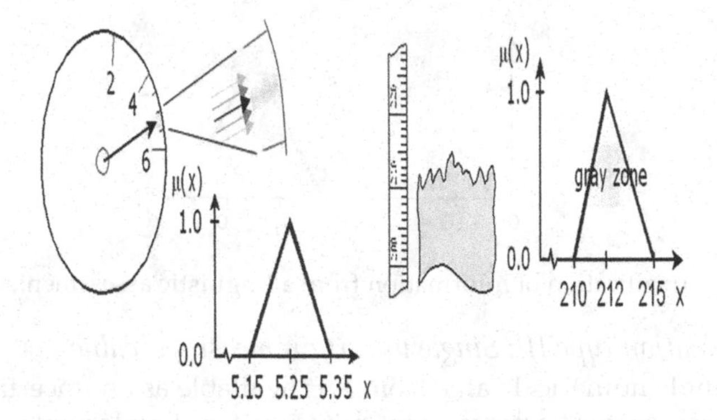


*Figure 11.5* Fuzzification of a single uncertain measurement due to imprecision of the measuring device or imprecise measuring point.

latter, are also assessed with membership values of $\mu_A(x) > 0$. This approach may be extended by selecting several crisp kernel sets for different membership levels (α-level sets) and by specifying the $\mu_A(x)$ in level increments.

## *Stochastic-fuzzy models*

Fuzzy randomness simultaneously describes objective and subjective information as a fuzzy set of possible probabilistic models over some range of imprecision. This generalized uncertainty model contains fuzziness and randomness as special cases.

Objective uncertainty in the form of observed/measured data is modeled as randomness, whereas subjective uncertainty (see Figure 11.6), e.g., due to a lack of trustworthiness or imprecision of measurement results, distribution parameters, environmental conditions, or the data sources, is described as fuzziness. The fuzzy-random model then combines but does not mix objectivity and subjectivity; these are separately visible at any time. It may be understood as an imprecise probabilistic model, which allows for simultaneously considering all possible probability models that are relevant to describing the problem.

The uncertainty model fuzzy randomness is particularly suitable for adequately quantifying uncertainty that comprises only some (incomplete, fragmentary) objective, data-based, randomly fluctuating information, which can simultaneously be dubious or imprecise and may additionally be amended by subjective, linguistic, expert-specified evaluations.

This generalized uncertainty model is capable of describing the whole range of uncertain information reaching from the special case of fuzziness to the special case of randomness. That is, it represents a viable model if the available information is too rich in content to be quantified as fuzziness without a loss in information but, on the other hand, cannot be quantified

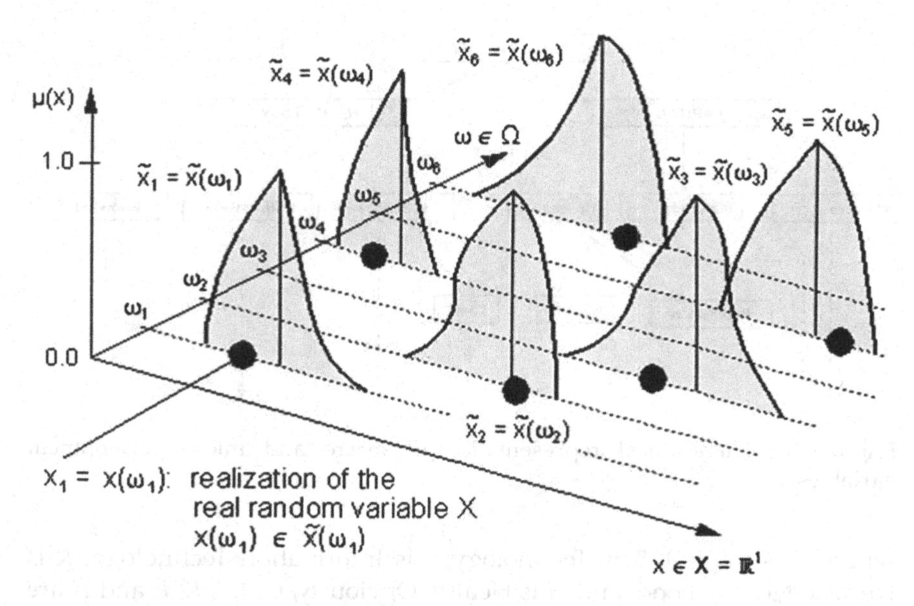


***Figure 11.6*** Model of a fuzzy-random variable.

as randomness due to imprecision, subjectivity, and non-satisfied requirements. This is probably the most common case in applied science.

Uncertainty quantification with fuzzy randomness represents an imprecise probabilistic modeling, which incorporates imprecise data as well as uncertain or imprecise subjective assessments in terms of probability. The quantification procedure is a combination of established methods from mathematical statistics for specifying the random part and of fuzzification methods for describing the fuzzy part of the uncertainty.

# Applications

## The development model

Development is the vital summation of all efforts made by man to increase the quality of life while sustainability is the continued successful upholding and enhancement of this quality of life by getting replenished the necessary ingredient/resources such as human labor and ecological resources. Development occurs when the intrinsic aspects are applied through technology to generate the physical aspect. Technology confirms the existence of the intrinsic aspects and creates the physical aspect to manifest development.

Considering development as a hierarchical (Figure 11.7) multivariate nonlinear function,

$$D = (C, T, I, K, F, H)$$

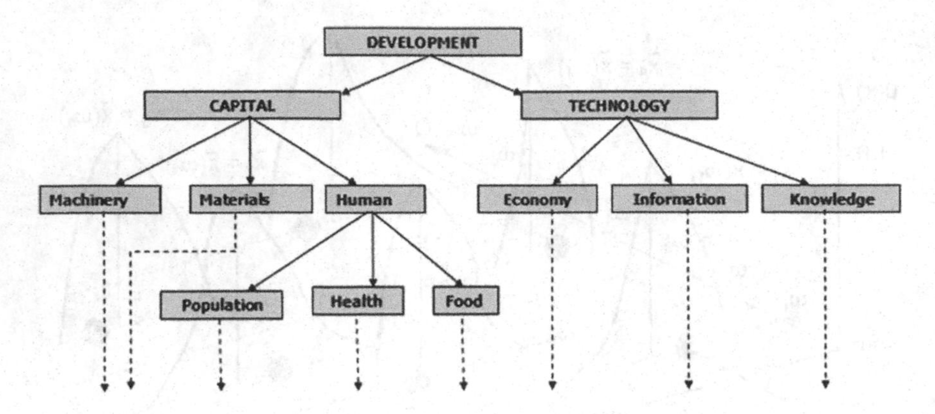


*Figure 11.7* Hierarchical representation of macro and micro development variables.

where $C$ is Capital; $T$ is Technology; $I$ is Information Technology; $K$ is Knowledge; $F$ is Food; and $H$ is Health; Obviously, $C$, $T$, $I$, $K$, $F$, and $H$ are not independent variables.

A compact analysis of the behavior of $D$ can be facilitated by the following dimensionality reduction.

Capital can be defined as an inventory of infrastructures machinery $m$, that is, plant, equipment, etc., materials $m_2$ and human resources $h$.

$$C = C(m_1, m_2, h)$$

Since human resources, $h$, itself can be viewed as function of population, health, and food, then the functional representation for $h$ alone becomes

$$h = h(P, H, F)$$

Thus, Capital $C$ can be represented as a compound functional representation as:

$$C = C\big(m_1, m_2, h(P, H, F)\big)$$

Following a similar argument, since the technology variable ($T$) depends on the average performance measure of all productive processes in the economy ($E$), the level of development and utilization of information technology ($I$), the average experience and intelligence in the society (i.e., knowledge, $K$), $T$ becomes

$$T = T(E, I, K)$$

Thus, the development nonlinear equation can be written as

$$D = D(C,T) = D\big(C(m_1, m_2, h(P,H,F)), T(E,I,K)\big)$$

This is obviously functionalizing development at a macro level. At the micro level are myriads of variables on which each of the variables above depends. The decision-making organelle in optimizing for development definitely consists of optimization at the macro and the micro level functions.

Uncertainty of different kinds influences decision-making and hence development. In optimizing functions of development, the effects of uncertainty cannot be ignored or rationalized; else, the results of such decision will at best be out of relevance to effect economic growth. At the micro level, where the numerous variables that influence decision-making at the macro level are cognizance must be taken of uncertainty in quantifying development variables. Inherent in the quantification are elements of uncertainty in terms of variability, imprecision, vagueness of description, randomness. As amply explained in earlier sections, such quantities are better realized as combinations of fuzzy and random variables.

For example, food security (see Table 11.1) is influenced by a myriad of variables, including level of mechanization of agriculture, population, incentives, soil conditions, climatic conditions, etc. These variables may not be measured precisely. Some may be fuzzy, some stochastic, and others manifesting a combination of stochasticity and fuzziness in their quantification. In effect, quantification of food ($F$) in the development model is a fuzzy-stochastic variable.

***Table 11.1*** A typical uncertain description of variables influencing food production

| Food-related variable | Description | Class of uncertainty |
|---|---|---|
| Mechanization | Linguistic description (e.g., level of mechanization) | Fuzzy |
| Population | A range of numbers, random over a time space | Fuzzy-stochastic |
| Incentives | A range of numbers | Fuzzy |
| Soil conditions | A random variable subject to variation over time | Stochastic |
| Climatic conditions | Quantitative and variable over time space | Fuzzy-stochastic |
| Yield | Quantitative description | Fuzzy-stochastic |

The other variables in the development above can also be viewed as fuzzy/stochastic variables, thus the Developmental model above can be viewed as a fuzzy-stochastic developmental model.

$$D = D\left(\overline{\overset{\sim}{C}}, \overline{\overset{\sim}{T}}\right) = D\left(C\left(\overline{\overset{\sim}{m}}_1, \overline{\overset{\sim}{m}}_2, h\left(\overline{\overset{\sim}{P}}, \overline{\overset{\sim}{H}}, \overline{\overset{\sim}{F}}\right)\right) T\left(\overline{\overset{\sim}{E}}, \overline{\overset{\sim}{I}}, \overline{\overset{\sim}{K}}\right)\right)$$

where the symbols "_" and "~" represent fuzzification and randomization, respectively, of the various variables.

## *The Optimization of the Fuzzy-Stochastic Development Model*

The realization of the relational effects of the variables of development is in hierarchies, even at the macro level. A good strategy for optimizing the function is to use the concept of multilevel optimization model.

Multilevel-optimizing models are employed to solve decentralized planning decision problems in which decisions are made at different hierarchical decision levels in a top-to-down fashion. Essentially the features of such multilevel planning organizations are that:

- Interactive decision-making units exist within a predominantly hierarchical structure.
- Execution of decisions is sequential, from top to lower level.
- Each unit independently maximizes its own net benefit but is affected by the actions of other units through externalities.
- The external effect of a Decision Maker's (DM's) problem can be reflected in both the objective function and the feasible decision space.

The mode of execution of such decision problem is that:

- The upper level DM sets his goal and accepts the independent decisions at the lower levels of the organization.
- The upper-level DM modifies it within the framework of the overall benefit of the organization.
- The upper level DM's action further constraint the lower level decision space and may or may not be acceptable at that level. If it is not acceptable, the upper level DM can still make a consensus that the constraints are relaxed further.
- This process is carried out until a satisfactory solution to all levels and units of decision-making is arrived at.

In solving the fuzzy-stochastic development model, each level of the macro-level model is taken as a decision level optimizing variables of its concern.

The complete fuzzy-stochastic multilevel formulation of the fuzzy-stochastic development model is, therefore:

$$\underset{\tilde{\tilde{C}},\tilde{\tilde{T}}}{\text{Max}}\,\tilde{\tilde{D}}$$

where, $\tilde{\tilde{C}},\tilde{\tilde{T}}$ is obtainable from

$$\underset{V_3}{\text{Max}}\,f(V_2)$$

with $V_2 = \left\{\tilde{\tilde{m}}_1, \tilde{\tilde{m}}_2, \tilde{\tilde{h}}, \tilde{\tilde{E}}, \tilde{\tilde{I}}, \tilde{\tilde{K}}\right\}$ which again is obtainable from

$$\underset{V_3}{\text{Max}}\,f(V_2)$$

where $V_2 = \left\{\tilde{\tilde{X}}, \tilde{\tilde{P}}, \tilde{\tilde{H}}, \tilde{\tilde{F}}\right\}$ and

$\tilde{\tilde{X}}$ = relevant variables at the micro-level of the development model.
Subject to:

$$\tilde{\tilde{C}},\tilde{\tilde{T}},v_2,v_3,\tilde{\tilde{X}} \in S$$

where $S$ is the constrained space of development variables (constrained by the limitations in realizing the various variables).

Realistically, this type of model may be difficult to solve for large spaces such as the national development pursuit but may be solved with smaller decision spaces. The worst solution scenario will be those that are not amenable to analytical solutions for which there are many heuristics to be coupled with simulation-optimization techniques.

## *Urban transit systems under uncertainty*

A relevant work (Ibidapo-Obe and Ogunwolu, 2004) on DSS under uncertainty is the investigation and characterization of combinations of effects of fuzzy and stochastic forms of uncertainty in urban transit time-scheduling. The results of the study vindicate the necessity for taking both comprehensive combinations of fuzzy and stochastic uncertainties as well as multi-stakeholders' objective interests into account in urban transit time-scheduling. It shows that transit time-scheduling performance objectives are better enhanced under fuzzy-stochastic and stochastic-fuzzy uncertainties and with the multi-stakeholders' objective formulations than with lopsided single stakeholder's interests under uncertainty. Figure 11.8 shows the modeling framework, which covers the following:

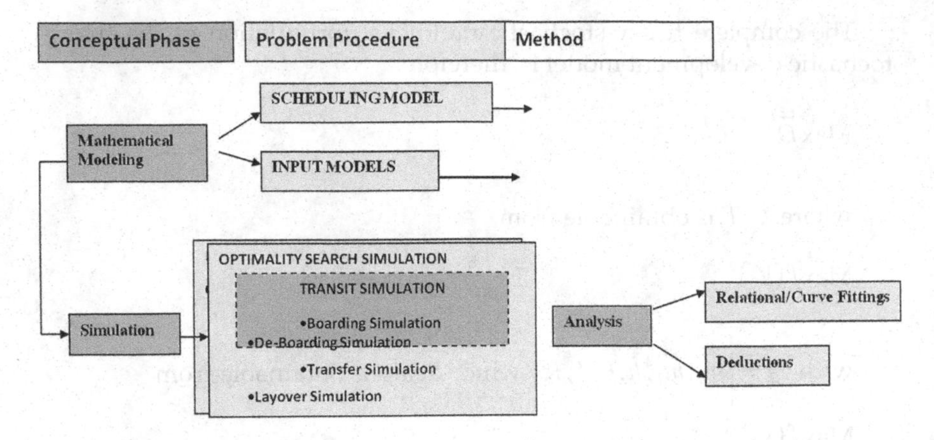


*Figure 11.8* Conceptual, procedural, and methodic frame of the transit time-scheduling research study.

- Mathematical modeling
- Algebraic, Fuzzy, and Stochastic Methods
- Multi-Objective Genetic Algorithm
- Algorithmic Computer Simulation Technique
- Mixed Integer Models
- Max-Min Techniques

### *Water resources management under uncertainty*

An on-going research on water resources management also suggests the need for incorporating comprehensive scientific uncertainty into the development of the models (Figure 11.9). This uses a three-level hierarchical system model for the purpose of optimal timing and sequencing of projects for water supply, for the optimal allocation of land, and for optimal irrigation. Major functions and inputs in the hierarchical fuzzy-stochastic dynamic programming developmental model are to be realized as fuzzy, stochastic, and or fuzzy-stochastic inputs of the model. Particularly, the demand and supply of water over a time horizon are taken as fuzzy-stochastic.

## *Energy planning and management under uncertainty*

One of the immediate future thrusts is on Energy Planning and Management Modeling under Uncertainty. Energy information systems are organized bodies of energy data often indicating energy sources and

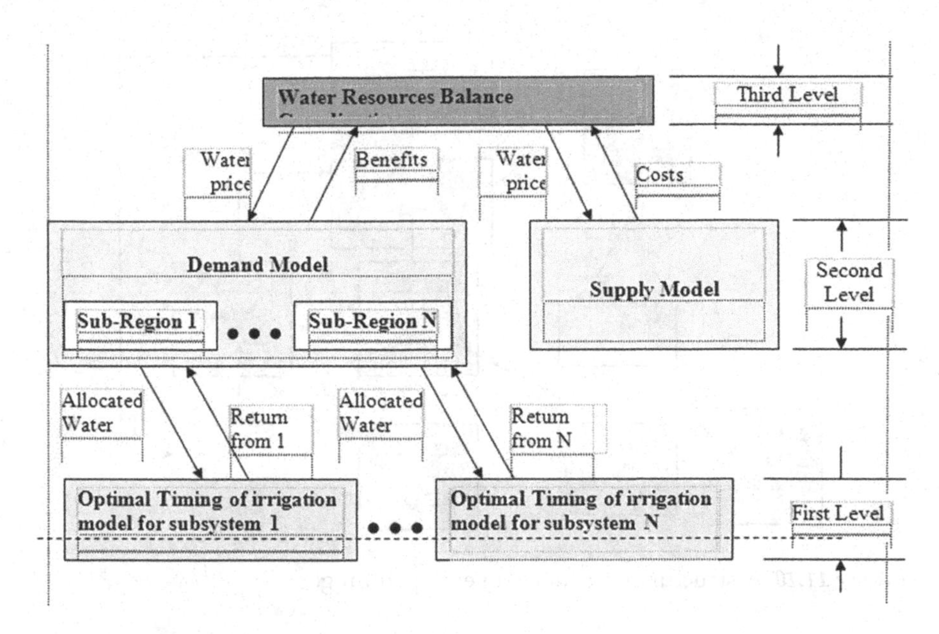


*Figure 11.9* Multilevel dynamic programming structure for planning and management of agricultural/water system.

uses with information on historical as well as current supply–demand patterns, which are naturally bugged with uncertainties (see Figure 11.10). These information systems draw upon energy surveys of various kinds as well as on other sources of information such as the national census, (another potential source of uncertainty in evaluation over time and space); information on energy resources and conversion technologies as well as consumption patterns (which are also better realized considering inherent uncertainties). A typical structure of national energy planning system is as shown below.

University Admissions Process in Nigeria: The Post-UTME (Unified Tertiary Matriculation Examination) Test Selection Saga.

There are many issues related to the Nigerian Education System (universal basic education [UBE], the 3–3 component of the 6–3–3–4 system, funding, higher education institution admissions, consistency of policies, etc.).

In spite of the opening up of the HEIs (Higher Educational Institutions) space to more states, Private, Open, and Trans-National Institutions; the ratio of available space to the number of eligible candidates is 87:1000 (University of Lagos 2006). It is therefore imperative that the selection mechanism be open and rational. The realization problem for rational access is further compounded with the introduction of Post-UME test. With two assessment scores per candidate given the Department/

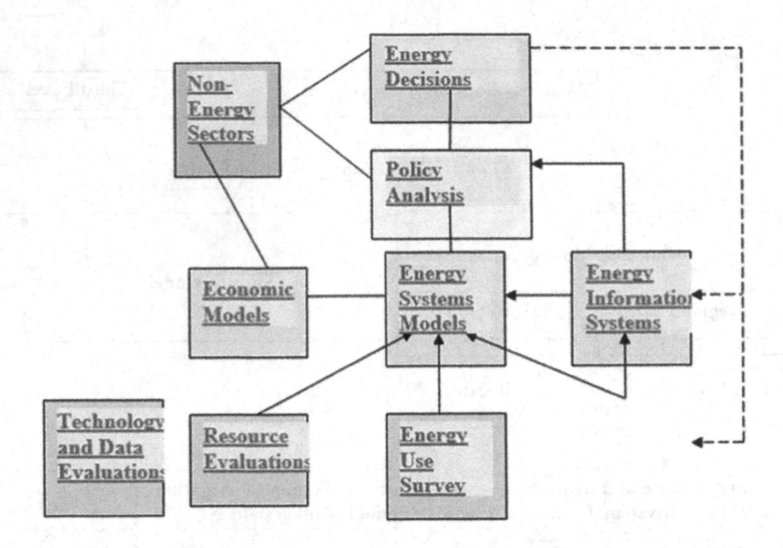


*Figure 11.10* A structure for national energy planning.

Faculty/University of choice quota, the question is how best can we optimize a candidate's opportunity to be admitted? That is:

- How do we establish that a particular candidate actually sat for the JAMB examination?
- Is the JAMB examination score enough to measure competence?
- Is it sufficient for the various Universities to select based on their own examinations?
- To what extent will the University Post-UME test help judge candidates' suitability?
- Could there be a rational middle-of-the-road approach in determining a candidate's admission to the University based on the two regimes of scores he holds?
- To what confidence level can a University assert that it has selected rationally?

Answers to these and many more questions and issues on this subject matter are subjects of wide variability, imprecision, and vagueness. The objective is to formulate a fuzzy-stochastic Decision Support Models for the resolution of this matter.

The just-concluded University of Lagos admission process was reflective of the underlying uncertainty in rationally giving candidates to its programs based on the two regimes of scores:

Stochastic Aspect: A candidate for admission must reach a particular scholastic level at the JAMB UME examination. UME is a highly stochastic scheme with randomness in performance over time and space. The various performance levels of all candidates for a particular Course/Faculty/University may be assumed as normally distributed.

Fuzzy aspect: Candidates for admission need another scholastic assessment based on the score of the Post-UME screening exercise conducted by the University. This is fuzzy based on a reference course Faculty/University mean score(s). The Post-UME screening score may be assumed as having membership grade distribution which is equivalent to the fuzzy set (represented by the Course/Faculty/University mean score (xi) to 100/400).

In other words, candidates must satisfy Post-UME screening (fuzzy) criterion before being considered for admission based on the UME scores (Stochastic).

Mathematically expressed,

$$\text{Max}\, A = A\left(PU_s, \frac{U_s}{q_o}\right)$$

subject to

$$q_o = q_o(\text{merit}, \text{catchment}, \text{ELDS})$$

where

| | | |
|---|---|---|
| A | = | Probability of admission at the University of First Choice and First Course |
| $U_s$ | = | UME score |
| $q_o$ | = | Course admission quota |
| $PU_s$ | = | Post-UME screening score |
| Merit | = | absolute performance (independent of state of origin of candidate) |
| Catchment | = | contingent states of location of institution |
| ELDS | = | educationally less disadvantaged states in the federation. |

Further exposition on the above discussions can be found in Beer (2004), Benjamin and Cornell (1970), Ibidapo-Obe (1996, 2006), Mamdani and Assilian (1975), Möller and Beer (2004), Ogunwolu (2005), Olunloyo (2005), Rayward-Smith (1995), Rommelfanger (1988), Takagi and Sugeno (1985), Zadeh (1965, 1973) and Zimmermann (1992).

## Conclusions

This chapter outlined scientific uncertainty in relation to optimal DSS management for development. The two principal inherent forms of uncertainty (in data acquisition, realization, and processing), fuzziness and stochasticity as well as their combinations are introduced. Possible effects on developmental issues and the necessity for DSS which incorporate combined forms of this uncertainty are presented.

In particular, the major focus of this submission is the proposal to deal with scientific uncertainty inherent in developmental concerns.

Uncertainty should be accepted as a basic component of developmental decision-making at all levels, and thus, the quest is to correctly quantify uncertainty at both macro and micro levels of development. Second, expert and scientific DSS which enhance correct evaluation, analysis, and synthesis of uncertainty inherent in data management should be utilized at each level of developmental planning and execution.

## References

Beer, M. (2004). Uncertain structural design based on non-linear fuzzy analysis. Special Issue of ZAMM, *ZAMM-Journal of Applied Mathematics and Mechanics/ Zeitschrift für Angewandte Mathematik und Mechanik: Applied Mathematics and Mechanics* 84(10–11), 740–753.

Benjamin, J.R. and Cornell, C.A. (1970). *Probability, Statistics and Decision for Civil Engineers*. McGraw-Hill, New York.

Ibidapo-Obe, O. (1996). *Understanding Change Dynamics in a Stochastic Environment*. Inaugural Lecture Series. University of Lagos Press, Lagos.

Ibidapo-Obe, O. (2006). Modeling, identification/estimation in stochastic systems. In A.B. Badiru (Ed.), *Handbook of Industrial and Systems Engineering*, vol. 14. CRC Press/Taylor & Francis Group, New York, pp. 1–16.

Ibidapo-Obe, O. and Asaolu, S. (2006). Optimization problems in applied sciences: From classical through stochastic to intelligent metaheuristic approaches. In A.B. Badiru (Ed.), *Handbook of Industrial and Systems Engineering*. vol. 22. CRC Press/Taylor & Francis Group, New York, pp. 1–18.

Ibidapo-Obe, O. and Ogunwolu, L. (2004). Simulation and analysis of urban transit system under uncertainty – A fuzzy-stochastic approach. *Proceedings of the Practice and Theory of Automated Timetabling PATAT*, Pittsburg, PA.

Mamdani, E.H. and Assilian, S. (1975). An experiment in linguistic synthesis with a fuzzy logic controller, *International Journal of Man-Machine Studies* 7, 1–13.

Möller, B. and Beer, M. (2004). *Fuzzy Randomness – Uncertainty in Civil Engineering and Computational Mechanics*. Springer, Berlin.

Ogunwolu, F.O. (2005). Time-scheduling in urban transit systems: A multi-objective approach under fuzzy and stochastic conditions. *Ph.D. Thesis*, University of Lagos, Lagos, Nigeria.

Olunloyo, V.O.S. (2005). Project Nigeria and Technology. Nigerian National Merit Award Winner's Lecture.

Rayward-Smith, V.J. (1995). *Applications of Modern Heuristic Methods*. Alfred Walter Limited Publishers in association with UNICOM, Henley-on-Thames, Oxfordshire, England, pp. 145–156.

Rommelfanger, H. (1988). *Fuzzy Decision Support-System*. Springer, Berlin.

Takagi, T. and Sugeno, M. (1985). Fuzzy identification of systems and its applications to modeling and control, *IEEE Transactions on Systems, Man, and Cybernetics* 15, 116–132.

Zadeh, L.A. (1965). Fuzzy sets, *Information and Control* 8, 338–353.

Zadeh, L.A. (1973). Outline of a new approach to the analysis of complex systems and decision processes, *IEEE Transactions on Systems, Man, and Cybernetics* 1, 28–44.

Zimmermann, H. (1992). *Fuzzy Set Theory and its Applications.* Kluwer Academic Publishers, Boston, MA.

# Index

## M



## N



## O



## P



## Q



## R



## S

Printed in the United States
by Baker & Taylor Publisher Services

Printed in the United States
by Baker & Taylor Publisher Services